BUILDING A
SPECIAL

First published in March 2021

A catalogue record for this book is available from
the British Library.

ISBN 9 781 78521 705 0

Library of Congress control no. 2019957244

Published by J H Haynes & Co. Ltd.,
Sparkford, Yeovil, Somerset BA22 7JJ, UK.
Tel: 01963 440635
Int. tel: +44 1963 440635
Website: www.haynes.com

Haynes North America Inc.,
859 Lawrence Drive, Newbury Park,
California 91320, USA.

Printed in Malaysia.

Senior Commissioning Editor: Steve Rendle
Copy editor: Steve Rendle
Proof reader: Dean Rockett
Page design: James Robertson

Images
The vast majority of the photographs showing
the construction of the car were taken by Ant or
Chris Hill as the build progressed. John Lakey
also contributed a couple of photographs. All of
the amazing graphics were produced by Paul
Cameron. There are also number of stills taken from
the *Master Mechanic* television show. Any other
contributors are credited in the relevant captions –
thank you to all of them.

BUILDING A SPECIAL

Following the build of Ant's own
classic F1 single-seater special

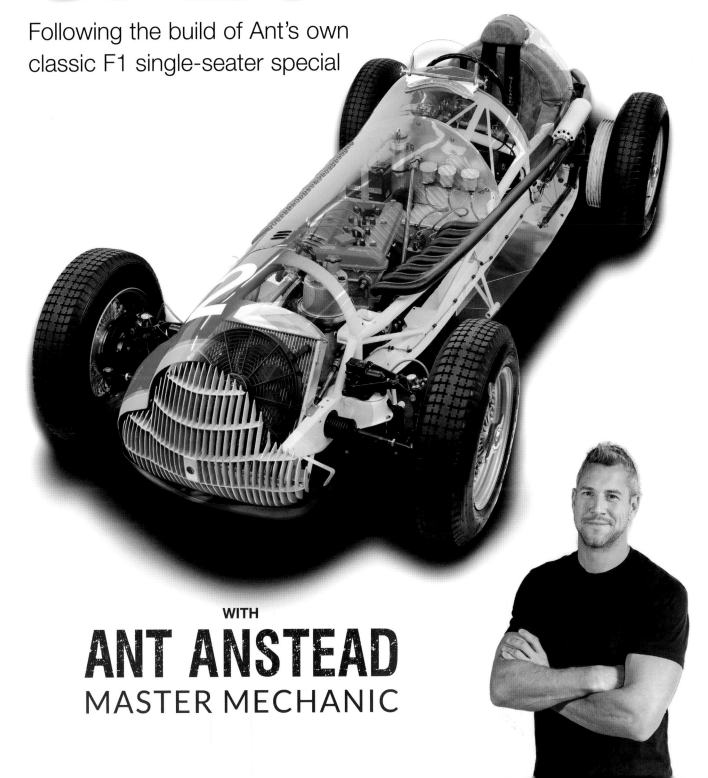

WITH

ANT ANSTEAD
MASTER MECHANIC

Contents

(Ash Sater)

Introduction

This isn't just a book. It might look like just a book, but it isn't. It's also a great journey. I wrote this while building the very car you see on the cover and I have documented the process here. If you were looking for a sign, this is it! This is your invitation to start your own journey – to build yourself a car.

Many great journeys start without necessarily knowing the precise route, and mine was no different, although I did know the destination. I had to change course a couple of times, navigate using obscure methods, and accept that things wouldn't always go to plan. Sometimes I got lost. Consider this book as the road map I drew with the benefit of knowing the destination. You don't have to follow it, but you're very welcome to do so.

So, what's so special about the Alfa 158? Simply, it's the most successful Formula 1 car ever. It was designed in the 1930s, and is a far cry from today's carbon-fibre rockets – it looks more like a fish than a spaceship. The 158 won the first-ever F1 race, way back in 1950 at Silverstone, and is cemented in history as one of the greats. Oh, and she is beautiful as well as fast.

Another motoring legend born in the 1930s was Mr John Haynes. Writing this book in a similar style to a Haynes Manual is a boyhood dream come true for me. Like so many young car fans, I grew up relying on Haynes. I never did have a teacher, I learned by doing, using books just like this. My own car-building journey started after finding John Haynes's booklet *Building a 750 Special*, which he wrote while still at school. I cut my teeth building Austin Seven specials, and have built multiple specials since. John Haynes's books inspired me, and now it's my turn to hopefully inspire more people to build cars.

This book is also a perfect companion for my Discovery television show, *Ant Anstead – Master Mechanic*, which documents the build of my Alfa 158 special over 10 episodes. A very small crew and myself not only built the car, but filmed the whole process and had great fun doing so. If you've got the book, then I suggest you catch with the show – the two are perfect companions.

You might want to sit down for this bit... Building cars can be, by its very nature, dangerous. You can hurt yourself on this journey. Cars are very heavy for a start. You'll be using power tools that create sharp edges and will spit red-hot bits of metal at you, and highly inflammable chemicals and fuel are used. Fortunately, there are ways to protect yourself from danger. Please use them. After 25 years of doing exactly this work, I have become a bit familiar and sometimes a bit slack, which you may notice in some of the photographs here. Forgive me – it's a classic case of please do as I say, rather than as I do!

That said, grab your protective goggles and gloves, roll down your sleeves while welding, and I hope you enjoy following my journey, as I build my version of an Alfa Romeo 158 Formula 1 car. If just one person out there is inspired to start their new journey and do the same, my life is complete!

▶ **Juan Manuel Fangio sitting in an Alfa Romeo 159 Alfetta. His car wore the number 22 when he clinched the championship by winning the Spanish GP on 28 October 1951, his first world title and Alfa's last. To me, this is the iconic image of the most heroic of drivers in the greatest of cars, wearing a number which has always fascinated me.** *(Centro Documentazione Alfa Romeo – Arese)*

(Ash Sater)

CHAPTER 1

Building something special

As far back as I can remember I have always had an overwhelming desire to build things. I needed to understand how things worked, but equally why they didn't. I was captivated by engineering problems and even more so by their solutions. And often it was the simplest of things that tickled me the most.

Like loads of young kids I started at an early age with click-together building blocks, graduating to making things from whatever I could find laying around. A cardboard box, an old radio, or even a discarded bathtub became useful items in my child's world of building. Playtime was easy, being one of four boys. Back then I leaned towards making things with wheels – I didn't just make a thing, I made a thing that moved. I never realised it at the time, but I was paving the way towards becoming a car builder.

At the age of 13, I built a go-kart using an old petrol lawnmower I bought (without telling my parents) from a local flea market. I think I paid £2 in total, bidding 50p at a time. I dragged the heavy lump home and set about removing its motor to place in a wooden 'rolling chassis' I had cobbled together. I added some stolen pram wheels and made steering from lengths of rope and bamboo garden canes. Think of the classic kids' go-kart, but unlike those of my peers, mine would have an engine!

I remember cutting every part of the angular, wedge-shaped body from chipboard that was hanging around my dad's garage. I think it came from a broken wardrobe. The hand saw I used was so blunt and rusty I had blisters across the palms of both hands building that 'car'! Worse still, I never did

▼ **From an early age, I just had to get moving, although this vehicle could do with some wheels!** *(Anstead Family)*

▼ **Was it just for my family in the '70s that clothing for kids seemed optional?** *(Anstead Family)*

get it running! Throughout my early teens, building improvised chariots (and recklessly throwing myself down hills) became a familiar theme.

My first car

At the age of 17, my world changed the moment I got my driving licence. I obsessed about getting on the road, and spent most of the months leading up to my birthday preparing for it. I passed my driving test first time. Freedom! My mates all had regular cars of that period like Metros, Fiestas, Minis, Astras, Golfs and so on (that we paid peanuts for back then, but are now worth a small fortune) – the regular car fodder of an English teenager. Unlike my friends however, I had already been bitten by 'the bug' and, of course, my first road car was a classic – a Vermillion Orange MG Midget... called Bridget. She had rusty wheel arches and a torn black roof with the rear screen held together by duct tape, and she rattled and leaked oil. If it rained, I got wet, and it seemed that in England back then, it always rained! I learned so much from that car. She required daily repairs to keep her running – she was trouble – but I absolutely loved her. I might go as far as to say Bridget was my first real girlfriend.

Having to continually maintain Bridget to keep her on the road fired a passion within me. I worked on her on an almost daily basis just to get around, and I loved every minute of it. I quickly worked my way through numerous classic cars, buying wrecks, fixing them and making some profit. Soon, the maintenance

of a classic wasn't enough for me. I needed a bigger challenge. I wanted to build a car from scratch – from the ground up – a new vehicle altogether.

A move to kit cars

I did my research, worked as much overtime as I could waiting tables at hotels and working in various school kitchens to raise some cash, and I purchased an off-the-shelf, purpose built 'kit car' made by a UK company called Tiger, based in Cambridgeshire.

I remember convincing my dad to take me to buy the 'donor'. A bright red Ford Sierra I found in the classifieds of a car magazine. I paid just £50. It had all the parts I needed – front and rear suspension, 1800 Pinto engine and a manual gearbox – perfect! When we arrived to pick it up, the car had lost half its exhaust and sounded like a broken war plane. So, in a moment of genius, it was decided that Dad would tow me and I would steer the Sierra the 80-odd miles home, which included a scary section of the M25. Probably not a great idea in hindsight.

It was dark as we pulled into our village, and 'snap'! The tow rope broke leaving Dad and me to push the car the final half-a-mile or so home. Those last few yards were hilarious, and hard. I am certain he thought I was crazy, but also chuffed to be helping. I knew in my mind how this car would eventually turn out. I just needed the Sierra in the garage to get started. Thank you Dad!

The Tiger went together relatively easily.

▲ **While all my friends drove cars of the period, like Minis, Golfs and Escorts, my first car was, of course, a classic – an MG Midget called Bridget. I adored her!**
(Ant)

▶ Ron Champion's original Haynes *Build Your Own Sports Car* book had a profound effect on me, and led me to start my first business.

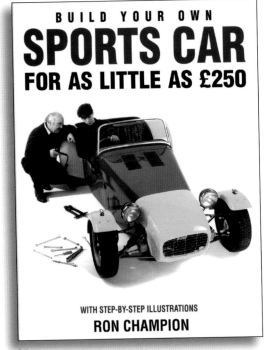

BUILD YOUR OWN
SPORTS CAR
FOR AS LITTLE AS £250

WITH STEP-BY-STEP ILLUSTRATIONS
RON CHAMPION

▼ Assembling a kit car is a great way to get into car building, with the bulk of the hard work done for you.
(Nick Wood)

Of course, at the time I didn't realise that the Tiger company had spent years on research and development, which made my first kit car experience a good one. I just thought building cars from scratch was simple, and 'boom', I'd caught the car-building bug!

After the Lotus 7-inspired Tiger, I stumbled upon a Haynes book by Ron Champion called *Build Your Own Sports Car*. Unlike a 'kit car', it taught home builders how to build their own car from scratch – detailing how to make all the components. From that book the 'Locost' craze was born. Literally hundreds of cars were built, and a one-make race series was developed to satisfy the appetite of the car's fans. Ron's brilliant step-by-step guide opened up a whole new world of engineering to me. Ron was my kind of person.

On the back of the popularity of Locost, I started my first business. There were literally hundreds of people building Ron's cars at home, and I worked out that most people were happy bolting things together, but not building things from scratch. I set about supplying those harder-to-make parts – from chassis and wishbones to fuel tanks and bodywork. I created a whole cottage industry supplying components to home-builders. From my dad's garage direct to Joe Bloggs. As word of mouth grew, so did the business.

That period was so pivotal to my understanding of the real car industry – supply, demand, what worked, what didn't. I built a black book of contacts and suppliers. Parts aren't just parts, they mean a lot to the person that receives them. They transcend being shapes and materials, to being organs in a living structure – the car. I suddenly had respect for

everyone's projects, not just my own. All car builders are the same, and different. I fell in love with other people's projects too, I attended trade shows and offered my wares and I went on to supply Locost parts for many years.

One stand-out build belonged to Declan McDonnell. He won the 750 Motor Club Locost racing title in 2006 and 2007. I had a huge hand in that car. Declan and I focused on developing the handling, which clearly worked. It's a period I enjoyed and look back on with great affection. I had grown up. I wasn't just a kid who chose an orange Bridget over a Ford Fiesta, I was part of a driver winning an actual race series. That blew my mind, it still does.

Learning the basics

I realise not everyone's childhood was spent building racing cars. Growing up, we always lived in school accommodation provided with my dad's job (he was a catering manager at boarding schools) and, luckily, in the few house moves I remember we always had a home with a garage. That alone made me the luckiest kid in town. I spent many hours in these tiny single-car garages making mistakes, figuring out engineering and working out all kinds of odd little things for myself. My dad's garage saw a steady flow of classic cars and side projects. It became my workshop, my sanctuary.

I never went to college or school to be a mechanic, I was simply hands-on, making mistake after mistake, learning as I went. I filled my spare time going to car shows and annoying several local garages offering my services for free. My dad, bless

him, could hardly change a wheel, yet alone work on a car. Yet he supported me and my projects every step of the way. As I couldn't afford to get in any specialist help, I would research and learn each and every part of the car myself. What I didn't know at the time was that I was doing my own in-house automotive apprenticeship. This laid the foundations for my entire career.

Of course, starting with kit cars (and every car since) I always found myself straying off-script. It became a regular occurrence, and it made each car unique as I started to think of alternative ways to tackle builds that were apparently 'predetermined'. I was always that guy who built the flat-pack without using the manual. One Christmas (aged around eight), my parents bought me a self-build cardboard model of a World War 2 airfield. We didn't have much money growing up, but that was a perfect gift for me. I finished it before the Queen's speech, and my dad noticed I never opened the instructions once. I simply relied on the picture on the packet. And that's how it went with cars as I grew up. Without realising it, I was building **my** cars **my** way – creating something unique, a one-off, something that was all mine.

Finishing these first cars became real milestones for me – I stood back, silent, shocked almost. Wow, I had done it, creating something from nothing – my babies. But the baby stage didn't last long. I always sold these cars within a matter of weeks to fund other projects. I was obsessed. I was a car builder, not a driver, and I wanted grease on my hands, not wind in my hair, so this love affair with building cars began. For me, it was never about the drive, it was always about the build.

▲ **Locost racing became a significant class of grass-roots motorsport, leading to me building cars and supplying parts for many teams.**
(Steve Williams/750MC)

▲ **I think it's fair to say my dad was pretty cool.** (Anstead Family)

A car for everyone

I get asked a lot about the best car to buy, after all, there are a lot of options out there, from sports cars to off-roaders, hatchbacks to hot-rods. There is a car to suit any palate or, as the saying goes, 'an ass for every seat'. Right now, we live in an interesting time for the classic-car enthusiast. The car market has never been so active, and classic cars that seemed attainable not so long ago are now out of reach for many. Imagine your first family car – what it was worth then, and what it's worth now? Gulp! Exactly! My dad's first car was a 1957 Morris Minor. He paid just £25 for her on his 18th birthday, then he stepped up to a 1963 Austin Mini, for which he paid £120! Both considered low-budget, modest cars. My parents honeymooned in that Mini, camping and driving all over Cornwall. It had a huge Mickey Mouse sticker on the driver's door and Minnie Mouse on the passenger door. I haven't yet worked out if that's awesome, or awful! Either way, both those cars, albeit humble at the time, created memories that last a lifetime, and both of those cars are worth so much more money now,

▶ **My dad paid £120 for a 1963 Austin Mini, which is worth a bit more than that today. If anyone knows where this one is, please let me know.** (Anstead Family)

◀ That Mini took my parents all over the country. I don't want to think about what might have happened in the tent. (Anstead Family)

simply because they are old and relatively rare. Are they worth the money today? You bet they are! The memories are an extra bonus. What I would give to find and restore those two old cars.

Unfortunately, the thrill of owning, maintaining, restoring and keeping a classic car is becoming a pipe dream to most, because the costs are just too great. It hurts many of us that the cheap old wrecks we grew up with are now financially out of reach. I shudder at the thought of the cars I have sold for pennies over the years. If only we had known.

There are so many car lovers out there like me, who just want to retreat to the garage and tinker with something, whether they get it running or not, and say: 'I did that, that car is **my** car'. It's because of this that ever since I was 17 years old I've always had a car project on the go.

Into business

At 18, I joined the police. It turned out that shift work was perfect for a part-time car builder, and I balanced fighting crime and building cars rather well. It even came in handy when attending broken-down vehicles, and my unit joked that my initials were AA. I have lost count of the number of cars I restored or scratch-built during this period. At the rear of my Hailey Lane home in Hertfordshire, we had a single garage in a block of three. Very soon, I had negotiated the use of the other two garages with my neighbours, and even had cars parked in front and to the side. I had outgrown my home, and it triggered a huge turning point for me. My hobby had got out of hand, and I was now looking in the local paper for dedicated garage space.

A stand-out moment for me was when I answered

an advert from a local farmer, Tom Pateman. He showed me space in a wooden shed, but it was underwhelming and way too small for my growing needs. Opposite the sheds were some run-down brick barns with cows in. I walked over and said: 'What I really need is something this sort of size'. There was a decent-sized barn – enough to house six cars, or the one cow that was in there – but it had no doors, just a waist-high steel gate, no windows, no power, no lights, no water and, of course, it was ankle-deep in cow dung. Ping! There and then I offered to restore that barn and turn it into a workshop if Tom relocated the cow and gave me a year rent-free. Despite Tom clearly thinking I was crazy, he agreed and the deal was done. Daisy had to find a new home, as I was moving in.

I spent over three years in that barn, building cars, knocking out parts, making whatever I could,

▼ Despite having zero car technical ability, my dad did own some cool wheels and bad trousers.
(Anstead Family)

► Reading John Haynes's very first publication inspired me to start thinking about designing and building cars my own way.

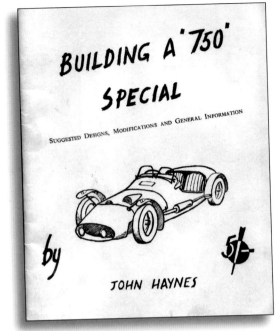

workshops over the years, and I even negotiated a workshop at a local football club as part of my professional playing contract, but that cow shed was the most significant for everything it represented. I had left the police to follow my car dreams, and I had become a parent, with early starts and late nights seven days a week. I was on my own and starting from scratch. I had no 'Plan B', and I launched a proper car company from that cow shed – I employed staff, built a team, launched new cars and eventually ended up building cars on TV because of that barn.

I'd spent years building what are called 'specials' – cars that have no specific heritage, design or origin – each one a unique vision of the builder's creative mind, but each one engineered for a specific purpose, whether that's a race car, a trials car, an off-roader or a weekend cruiser.

From specials to production cars

I remember getting my hands on the very first publication ever written by John Haynes, *Building a 750 Special*. The impact of that book cannot be underestimated. It not only laid the foundations for the Haynes dynasty of manuals that went on to help millions of home mechanics all over the world, it also helped me on my path as a bespoke car builder. It made me feel justified as a special builder. John Haynes, in one flappy pamphlet, had answered the nagging question in my mind. Yes, you can just use various car parts and build whatever you want. My build, my rules, my way. I was now an artist, and it was OK.

and slowly banking some profits. I did numerous restorations and built some really cool one-off machines. I created a proper little business. I bought second-hand tools when I could, and slowly built myself a workshop space to call my own. The winters were tough, and I would often wear three or four layers to stay warm. There was no heating, no water or toilet, and I was sick several times, but it was brilliant looking back.

I have used numerous barns, lock-ups and

▼ Building a special with so much heritage and history was a real honour. *(Michael Scott)*

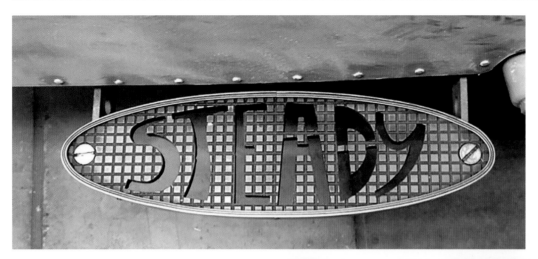

◀ STEADY. The side step I added to the famous *Steady Special*. *(Michael Scott)*

Just like John Haynes, I cut my teeth building Austin 7 based specials, retaining the chassis and running gear and making the rest up from scratch. I've lost count of the number of specials I have made over the years. I used mainly ladder chassis from early Austins or Morrises, but I also made chassis from scratch and created weird machines. A single-seater, Austin 7-based car with an offset rear axle that I built for the 'Wall of Death' stands out as a pretty cool one, and I still see that car running the wall today. I even had a hand in the reincarnation of the famous *Steady Special* and, yes, I am 100 per cent responsible for those low front headlamps, which (like the rear wheel cover) were hand-spun in aluminium. I even added the lovely little 'Steady' side-step last-minute – there were no rules, and it became my happy place.

After years of building one-off cars (basically making it up as I went along), I eventually graduated from 'have-a-go' specials to building a substantial car business. I have since built numerous cars from scratch, after many years of success with re-creations. Every part of these cars, from the chassis to the seats, is bespoke.

I then naturally graduated into designing my own road-legal, limited-production cars from scratch, drawing influence from the lines of iconic 1950s/60s GT cars. These cars evolved from ideas in my head, suddenly coming to life.

I still have a huge sense of pride when I see a car I designed and built driving on the public road, with a chassis number, log book and number plate. These are, after all, my babies – I was doing what companies such as VW, Ford and Nissan were doing – I was manufacturing.

Back to my roots

In this book, I am going back to my original roots, I'm going to build myself a special and I am going to take you through my build. Starting with nothing but a (fairly) clean workshop and a concept in my mind, I will, step-by-step, build myself a car that evokes all the passion and learning that makes me love this

unique sector of motoring so much. Nothing I do is rocket science – some parts may seem ambitious and beyond the scope of the home-builder, but there is a world of support out there, and I assure you it's a world of willing helpers. When you ask around, and let people know you are building a special you'll realise how many friends and family are interested, so don't be afraid to ask for help, or to involve them in your project.

The car I'm building in this book is intended for weekend events and track days, but eventually I will drive her on the public roads. I'll also give you some tips on how to make a special road legal, but all that is merely a bonus. This book is about the garage bit, the *build* itself, but it is simply a guide. Let me repeat that – THIS IS SIMPLY A GUIDE! It follows my build, for the *Master Mechanic* TV series, and is designed to inform you about the tools, components, materials, time and skills you need to build the same, or a similar car, for yourself.

I am not suggesting that you copy my car bolt-for-bolt, step-by-step – in fact I'm saying quite the opposite. Use the information in this book as a basis to build yourself something unique, something different, something that represents you.

Go and build yourself a SPECIAL!

▲ My build was followed for TV and can be viewed online. Go to the MotorTrend app and search for 'Ant Anstead Master Mechanic'.

CHAPTER 2

When is something a 'special'?

▲ One of the most important of the pre-war specials was the wonderfully named Multi-Union, seen here being driven by its creator, Chris Staniland, at Brooklands on 7 August 1939. The car was based on an Alfa Romeo Type B Monoposto, with a body built by J. S. 'Woolly' Worters. It is one of only three cars to lap Brooklands at over 140mph.
(Klemantaski Collection)

▶ *Specials* by John Bolster. He might well have defined for the first time what enthusiasts had been doing for centuries.

To help define what a special is, I must turn to the doyen of special builders and the author of the definitive book on the subject – the 1949 volume *Specials* – Mr John Vary Bolster.

On page one he states:

"A Special is a car built for a specific purpose by an amateur either entirely to his own design, or by combining the essential parts of a number of makes. The reason for building it is simply to produce a car with a better performance than anything the constructor could hope to afford to buy ready made."

However, special building is surely as old as the invention of the wheel itself. You can bet your baffled sump that Roman chariot racers had some secret tweak to make their chariots faster around Circus Maximus; maybe they discovered negative camber, or grease from a particular dead animal that made their axles run smoother... Who knows? The bottom line is, however, that humans will always want to improve performance by whatever parameter it is measured – it's human nature. A fair few of them will be skint though, so will try to achieve that for as little cash as they can. That desire, creativity and sheer fascination with machinery, is where the tradition of special building was founded, and it's a tradition I'm proud to continue in my own little way.

▶ **Red Bull Drift Brothers' Nissan S13 (Dark White) was King of Europe Champion in 2017 and is driven by Johannes Hountondji. The car features a GM Performance LSX 7.4-litre V8, with 620bhp and 860Nm of torque. The front styling has been taken from a Mitsubishi Galant, while the rear is almost like an enclosed pick-up bed, and the styling package is finished with a Rocket Bunny bodykit.** *(Red Bull Drift Brothers)*

Not every modified car – even a really heavily modified one – is a special though, because in car culture the word 'special' has come to have a sort of special meaning!

Those amazing drift cars, which mix and match components from Nissan, Toyota, Mitsubishi; the awesome Roger Clark Motorsport Subaru-based 'Gobstoppers', Andy Frost's frankly scary 3,650bhp (yes, you did read that correctly) Red Victor, and many other cars of a similar ilk are all special cars and amazing bits of engineering, but they are not, in this context, 'specials'. Neither are the 'resto-mods', which have become justifiably popular in recent years, such as the Eagle E-type, Singer Porsche or Frontline MGB.

▶ **The Roger Clark Motorsport Subaru 'Gobstopper' dominated the Time Attack Series in 2008 and 2009. Its 2-litre engine produced 875bhp@7,600rpm! It used a graduated nitrous oxide system by The Wizards of NOS. 0–60mph took 2 seconds. I want it!** *(Roger Clark Motorsport)*

◀ **My old mate Phil Glenister at the London Classic Car Show enjoying the poster car of the resto-mod scene, the Eagle Speedster; still recognisably a Jaguar E-type that's been tuned up to 11.** *(London Classic Car Show)*

◀ **Midlands-based Andy Frost has devoted a lifetime to building the fastest street-legal dragster in the world, and making it look like a Vauxhall Victor. At the time of writing it has 3,650bhp, and does a 6.02sec standing ¼-mile with a terminal speed of 251mph.** *(Andy Frost)*

The tradition I'm talking about really started after 1920, when World War 1 had finished, and Britain was suddenly flooded with cheap war-surplus machinery; trucks, planes, bikes and staff cars. The government needed money and needed to kick-start commerce, so had a massive yard sale, which boosted the economy by giving business cheap transportation and brought in cash. That, combined with a large number of demobilised men who'd been trained on various machines, as aero-engine fitters, tank drivers, despatch riders and the like, meant that the raw ingredients were all there. Very quickly the horse went from being the backbone of Britain's transportation network to a leisure activity for the few, as petrol-powered vehicles were seen as cleaner and greener than roads awash with horse manure. The result of all these available machines, and men who had been trained in some way to use them, was what would now be called a 'scene', from which grew the tradition of special building.

The special is about the 'have-a-go' ethos – such as builders in sheds with machines that are often built for competition. It's as much about the quantity of tea drunk as it is about horsepower. The spirit of taking part is more important than all that trivial competitive nonsense. It's a very British tradition that comes from the pre-war Brooklands and British hill-climb scene – Shelsley Walsh, Prescott and other tracks of that nature – featuring cars built on a separate chassis, and it's a car in that tradition that I am setting out to build.

That philosophy carries on to this day in the Vintage Sports Car Club (VSCC), an organisation that actually has a committee that decides what is and isn't a VSCC-eligible special. The VSCC insists on period components and construction methods: it even has a sentence in its rule book defining what constitutes a VSCC Special: *"Cars with major modifications or comprising components from a variety of eligible cars"*.

The key word there is eligible. Turn up with a modern fuel-injected Chevrolet LS3 V8 engine in a Model T chassis and they will be incredibly polite, but you will not have a VSCC-eligible special, though you would likely have a resto-mod (that creates a whole different conversation).

The car I'm building in this book wouldn't fit into the strict VSCC tradition, but I like to think it embodies the spirit of both the VSCC and John Bolster... "a car built for a specific purpose by an amateur either entirely to his own design, or by combining the essential parts of a number of makes". Yes, that's what I'm after here. The word amateur, in the best possible sense, sums the subject up perfectly for me.

Specials to F1

That cheerful, carefree and very British tradition of specials, established at hill climbs and trials in the years after what Bolster wonderfully calls in his book "the Kaiser business", also had a hand in creating the world-leading position Britain now has in motorsport engineering and F1.

Car sales took off in the 1920s, and the Austin 7 democratised British motoring, with over 290,000 built between 1922 and 1939. This created a ready availability of cheap Austin 7s just after World War 2, and with events being organised by the 750 Motor Club and others, in which specials based on Austin 7s could compete, the basis for a second post-war specials movement was in place.

Colin Chapman's first car – the Lotus Mk1 (as it retrospectively became known) – was a special based on a 1930 Austin 7, that he built in 1948 while he was still a university student, to go trialling with his girlfriend (and later wife) Hazel Williams. Within ten years he had stunned the world with the simply gorgeous and super-efficient Elite, which turned Lotus into a proper road-car manufacturer, and he had also become a regular entrant in F1. By 1960, a Lotus had won its first F1 race (Stirling Moss at Monaco) and by 1963 Lotus-Climax had won both the F1 Drivers' and Constructors' World Championships with Jim Clark. That's a long way to travel in 15 years, and of course he went on to dominate F1 and introduced more innovations than probably any other manufacturer ever has, or perhaps ever will. Famously, and perhaps ironically, he entered his cars in F1 for a number of years with 'John Player Special' title sponsorship.

▼ Colin Chapman's first car, a special based on a 1930 Austin 7, may not look like the first step to world domination, but it was. Here, the car is shown with Colin and Hazel Williams (who later became Mrs Hazel Chapman) in the passenger seat, competing in a trial. *(Colin Chapman Foundation)*

OX 9292

John Cooper's first car was also a special, built at home in 1935 with his father Charles Cooper when John was just 12. Charles also turned an Austin 7 into a special to race at Brooklands in 1936. In 1946, father and son built a special to compete in 500cc racing – a JAP-engined car, which used two Fiat Topolino front chassis and suspension assemblies welded back-to-back, giving independent suspension for all four wheels. Cooper grew into constructors, selling racing cars, and were the first team to win an F1 race with a rear-engined car, when Stirling Moss won in Argentina in 1958. In 1959 and 1960, Cooper won both drivers' and constructors' championships with Jack Brabham, and a little later had a go at making their friend and former special-building rival Alec Issigonis's new Mini go a bit faster. I think we know how that went...

Today, seven of the ten F1 teams have bases in 'Motorsport Valley' (centred on the Thames Valley in the West Midlands and Oxfordshire, UK), and those that don't source both talent and components from the UK. British drivers have won more F1 world championships than any other nation, and the UK motorsport industry is estimated to be worth over £10.5 billion. It employs over 45,000 people, exports over 75 per cent of its output, and is larger than the equivalent sectors in Germany, Italy and France combined. That may seem a long way from special building, but it's arisen from that 'can-do' attitude fostered by Bolster and his chums in the 1920s and taken on by Chapman, Cooper and many others in the late 1940s.

Special building is part of many aspects of British motoring history. In the following pages of this chapter, we look at a few examples of special building which, in hindsight, were much more significant than they might at first have appeared.

▲ **John Cooper, aged 12, in the Austin 7 Special with father Charlie Cooper in the centre behind.**
(Mike Cooper)

Bloody Mary – the embodiment of the special-building spirit

The most famous special in the world – *Bloody Mary* – was built and raced by John Bolster. It used an ash chassis and, eventually, two JAP engines connected together.

We can't talk about specials – especially the sub-

◄ **John Bolster driving *Bloody Mary* at Croydon Aerodrome, 5 September 1937. He was always spectacular...**
(NMM Beaulieu)

▲ ...but not always safe!
(NMM Beaulieu)

species known as the 'Shelsley Specials' – without looking at *Bloody Mary*. This iconic car is important firstly because it was built by John Bolster who, as mentioned previously, quite literally wrote the definitive book on the subject, secondly because it's got a great name, and thirdly because it has, to modern eyes, a charmingly bonkers element to its actual construction, and therefore sums up the spirit of the specials tradition.

John Bolster and his brother, Richard, built *Bloody Mary* in 1929 when both were still at school, as John later quipped: *"...with the object of driving around a field as dangerously as possible"*. Well, I think it's fair to say they achieved that aim, even with the 13bhp engine originally fitted!

The chassis consisted of three stout pieces of ash (the wood of choice for car makers before plywood gained some traction), one of which ran up the centreline, with all three joined together by two steel crossmembers. Much later in the car's life, these crossmembers were triangulated with gusset plates. In what initially seems like a perverse attempt to keep weight distribution low (but is actually quite logical when you consider the material weights), the three lengths of ash were mounted moderately high, with components hung off or below them, so the driver sat below the chassis with his or her derriere only 5in off the ground on a seat suspended from the centre and right main ash rails. The ash chassis lengths widened out at the rear, but the running gear all sat between the central and left rails.

Bloody Mary was built entirely with hand tools, and holes were drilled with a hand drill, which Bolster called the 'gut buster'. The car evolved massively throughout its life, while retaining its diminutive size and basic layout. The components really were culled from what could be found lying

around. Bolster, for instance, claimed that he never knew the origin of the steering box, but that it required one, yes **one** complete turn of the steering wheel from full-left to full-right lock. That was one of many reasons why the car was apparently more difficult to drive in a straight line than when cornering! In fact, after a scary moment, Richard Bolster stopped competing in *Bloody Mary* and built his own, slightly more conventional special which, confusingly for historians, used the same number plate as *Bloody Mary*. Bolster claimed this was inconvenient, as it meant they couldn't drive in close proximity... As a proud ex-police officer, I shall take a sideways look at that!

However, John continued to develop *Bloody Mary* and initially concentrated on fitting ever-more-powerful JAP engines. The first 760cc JAP nearly killed him when: 'The rear cylinder flew past my ear one day, followed by pieces of piston and sundry bits of hot metal'! The exclamation mark is mine... We all sometimes moan about the health-and-safety culture, but maybe someone needed to at least invent the concept. The car's engine development culminated in the use of a 1924 981cc four-cam side-valve JAP twin which, when tuned and run on alcohol fuel (Bolster claimed: '*Bloody Mary* seems to enjoy her alcohol as much as her owner'!) gave almost 50bhp which, in a car that weighed around 230kg, gave him many class victories and a few 'FTDs' (Fastest Time of the Day) in speed trials and hill climbs of the early 1930s. He improved the car by replacing the beautifully named Juckes gearbox with a four-speed unit made by Sturmey Archer, which – and he was quite affronted by this – cost him £7!

In search of more overall wins, Bolster considered supercharging *Bloody Mary* for the 1933

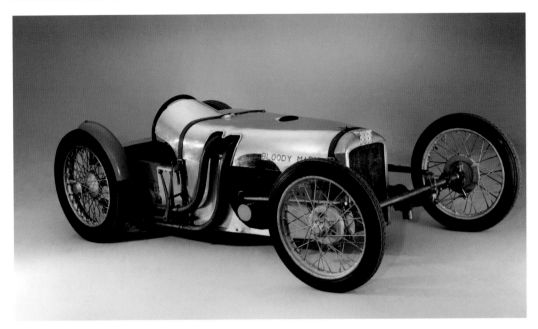

◄ Fittingly, *Bloody Mary* is now on display at National Motor Museum Beaulieu for all to see and is fired up occasionally and demonstrated. *(NMM Beaulieu)*

season, but decided the cost and the risk of engine explosions wasn't practical, so he bought another identical JAP engine and connected them together, as you do! The result was the definitive *Bloody Mary*, which was very successful up until the end of 1937. For 1938, Bolster built a new special using *Bloody Mary*'s two engines, plus another two connected together, but he never bonded with this car in the way he had with the original, so he returned to campaigning the original *Bloody Mary* after World War 2. He actually held the VSCC course record at Prescott Hill Climb between 1948 and 1953.

As for the origins of the car's name, restorer Sandy Skinner thinks it was because the two youngsters wanted to gently lampoon the attitudes of the time and says: 'They did it to taunt the commentators at race meetings, who would struggle to come up with euphemisms like "The Bolster Mary" or even "The Tudor Queen".' What a top chap! I'd love to have met John Bolster, and would like to think he would have approved of the project in this book. I'm sure he would certainly have jumped in and contributed to it.

1925 Morris Garages Kimber Special – The Octagon's origins

Out of all the great marques that fell into the melting pot of British Leyland in the 1970s, few have survived. Jaguar and Land Rover have become one company, but the only other remaining name is MG. Here is not the place to cover the whole history of MG; however I do think it's worth pointing out that MG originated from special building, so basing my car on one of MG's products has a germ of historic relevance!

William Morris founded Morris Garages in 1913 to sell his cars. In 1922, one Cecil Kimber became general manager of that business. Although Morris

was not interested in motorsport, Kimber was, and started building small numbers of more sporting Morris-based cars. By 1924, this experiment had become a small, but profitable, sub-business and Kimber had started using the MG – Morris Garages – name for his cars. Morris didn't stand in his way, because he could see this was a small, but nevertheless significant and profitable, market for his products, which was also potentially good for the company's image.

Although it wasn't the first car to carry an MG badge, most consider the first purpose-built full MG car to be, 'Old Number One', a special built to compete in the 1925 Lands End Trial. It utilised a

▼ Cecil Kimber at the wheel of MG 'Old Number One' (as it later became known), with his passenger Wilfred Matthews ascending Blue Hills Mine during the 1925 Lands End Trial. *(NMM Beaulieu)*

▲ Michael Scott, long time owner of the car, driving the *Steady Special* on the Grand Avenue at the London Classic Car Show. *(London Classic Car Show)*

Steady Special – a car I helped to reincarnate

The *Steady Special* is a Lancia Astura-based special designed and initially built by the doyen of motoring journalists, the late Ronald 'Steady' Barker.

I've had the pleasure of doing some work on this beautiful and unique car and was honoured to do so, because I grew up reading Steady's wonderfully irreverent musings on motoring in *CAR* magazine. The car started life as a 1934 Lancia Astura 3-litre V8 six-seater limousine, which was used by BOAC's director of engineering, and then by Brigadier General Herbert William Studd, but was acquired in the late 1940s by Steady for £75. He shortened its chassis by nearly 4ft, then fitted a sporting body, which used an Aston Martin DB2 clamshell bonnet. Steady rebuilt the engine, and in bored-out and tuned specification it produced almost 100bhp. He raced the car in this form very successfully in the early 1950s, achieving fastest laps at VSCC meetings at Silverstone, Oulton Park and Prescott.

A life-long Lancia enthusiast, Steady sold the car in 1955 to a then 18-year-old friend, Michael Scott, who used it extensively before selling it in 1976. Steady bought back the bodyless 'skeleton' in the mid-1980s, and sketched an elegant body design which kept the car visually and stylistically linked to the era in which the original Astura base car was created, a philosophy known in VSCC terms as PVT (Post Vintage Thoroughbred).

Steady made only modest progress with this until 2011, when Scott persuaded him to part with the unfinished project for a second time. Scott worked with renowned specialists Traction-Seabert and others to progress the build of Steady's design. Michael then handed the project to me to complete. I changed the front end, incorporating the lights into the body at a low level, altering the face of the car.

So there you have it – a 1934 V8-powered Lancia 'limo' re-imagined as a fast PVT tourer and ready for the next 85 years – the spirit of special-building at its very best.

The reincarnated *Steady Special* made its public debut in its new form on the Grand Avenue at the 2018 London Classic Car Show.

Michael Scott has put a huge amount of time, money and effort into finishing the *Steady Special* as a tribute to his lifelong friend Steady, who passed away in 2015, and has been rewarded with a car which is both beautiful and engaging to drive.

Building inspiration

The special-building community is vast and varied, and the choices available for my build were endless. The point of a special is to build something you want, and draw inspiration from anywhere you choose. For the build I'm describing in this book, I was inspired by one specific genre, and one specific car – 1950s F1 and the Alfa 158.

▼ Steady Barker's sketch of a body for the short-chassis Astura Special that he hoped would be acceptable to the VSCC as a PVT (Post Vintage Thoroughbred) Tourer. *(Michael Scott)*

Morris Cowley chassis, the rear section of which was modified by cutting off the frame and fitting upswept rails that went over the rear axle to secure splayed rear springs. The engine was a special OHV 1,548cc Hotchkiss unit with machined and polished ports. The special lightweight body was designed and built by Carbodies of Coventry for Kimber, and the final assembly of the car took place at the Morris factory. Kimber competed very successfully in the Lands End Trial that year, winning a gold medal in the Light Car Class. The road to building what was at one time the best-selling sports car in the world, the MGB, and my own first car, the MG Midget, really began with this special.

CHAPTER 3

The Alfa 158 story

(Centro Documentazione Alfa Romeo – Arese)

There are two motoring eras that really capture my heart – the 1930s and the 1950/60s. The '50s and '60s, of course, was the period of those glorious sports and GT cars, a period where coachbuilders and designers just seemed to get it right; the Ferrari 250SWB and GTO, Aston Martin DB4 GT and Zagato, Maserati 3500, Jaguar C, D and E-types, Iso Rivolta and Grifo, Facel Vega HK500, Austin Healey 3000, Bizzarrini 5300GT, AC Cobra, *and* the Ford GT40. The list is endless, and these cars are timeless, elegant masterpieces. I adore them, and have done since childhood. In fact that list could fill the whole book, but you get the idea... It's the reason why I love events like the Goodwood Revival, where you can still see many of these cars racing wheel-to-wheel.

▼ A whole field of GT40s, never an easy car to drive in any conditions, throwing up spray as they race into the sunset at the Goodwood Revival – sights, sounds and smells you don't quite get anywhere else. *(John Lakey)*

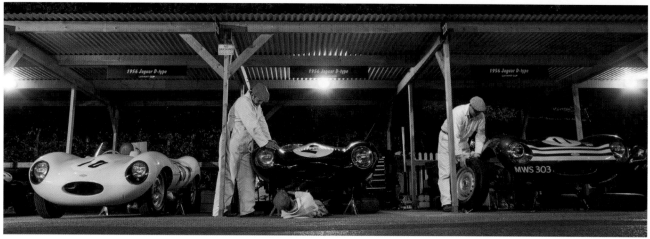

▲ The Goodwood Revival pits are a hive of activity late into the night, with mechanics and valeters making sure the cars look and sound perfect for the following day. John Lakey has made a speciality of shooting studio-style images in the night-time Revival pits, and I couldn't resist using a couple. *(John Lakey)*

Although I fell in love with the GTs as child (who doesn't love an E-type when they are 12?), as I got more involved with historic cars and historic motorsport it was the boat-tailed single seaters that began to captivate me. These were amazing cars, driven by heroes.

For me, there is one car from that era which towers above all others. It raced very successfully before and after World War 2, yet only a handful were built. A genuine one, should it ever come up for sale, would probably be valued somewhere between $20 million and $30 million, so I'll never own a real one. So, I'm compelled to build my own version of my dream

▶ Winner Luigi Fagioli in his Mercedes-Benz W25B overtaking Luigi 'Gigi' Soffietti in a Maserati 8CM as they enter the gasworks hairpin during the Monaco Grand Prix in 1935. *(Daimler AG)*

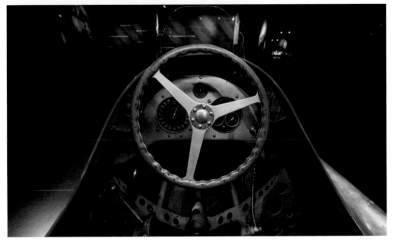

◀▼ **It's not just the compelling history of the Alfa 158 that I love, it's the unwavering beauty. There is not a single thing that I would change when I look at it.** *(Centro Documentazione Alfa Romeo – Arese)*

car, my way, from a more humble perspective. Yet, in some ways that's appropriate, as several F1 teams grew from special-building roots, so why not turn that around and build a special inspired by F1?

My special build is my homage to that iconic car, the Alfa Romeo 158.

Why does the 158 stand out to me above the other cars of the era? Well, first of all it's a visceral response to the look of it, I mean, just drink it in and then tell me it's not perfect in every single way? The overall outline and the design details; there's nothing you'd want to change, is there? For me, as a car builder, there actually aren't that many cars, of any era, that I can say that about. In fact, it's in my DNA to want to change cars – I've spent a lifetime doing exactly that!

It's more than that though – the story of the 158 is utterly remarkable. Its history is as compelling as its beauty, in terms of statistics, political intrigue,

and the character of the heroes who drove it: Varzi, Sommer, Farina, Fangio and many others.

To me, however, it is arguably the greatest racing car ever made, because nothing else achieved so much and remained so dominant for such a long period of time. Oh, and it's an Alfa, and it's red – very red – which is never a bad thing!

▶ **I had the chance to actually interact with a real Alfa Romeo 158 courtesy of owner Judy Giddings. Somewhat to my surprise I actually fitted, so I could drive it. Since I realised this I have been nice, very nice in fact, to Judy...** *(Master Mechanic)*

ALFA ROMEO 158 FACTS

■ The first two F1 World Champions – Farina (1950) and Fangio (1951) won their titles driving the Alfa Romeo 158/159 (the 159 was an updated 158 introduced for 1951), when the cars were already well over a decade old!

■ The car won its first race, on 7 August 1938, driven by Emilio Villoresi

■ It won its last F1 race, on 28 October 1951, driven by Juan Manuel Fangio

■ Between 21 July 1946 and 14 July 1951, a 158 won every race it was entered in. That is domination, and the design was eight years old when that domination *began*!

■ The car is the beginning of the Ferrari story. It was developed with, but not by, Enzo Ferrari, when he worked for Alfa, and it was at least partially his idea. This was the first car developed at Scuderia Ferrari. It was then taken away from Ferrari, as Alfa took back control of its motorsport department. Shortly afterwards, Enzo left the company.

■ The car started off with 195bhp in 1938, and ended its racing life in 1951 with 425bhp! That increase in power was achieved using the original engine castings! Both figures are remarkable for a 1.5-litre car of that era.

■ Lastly, and for me this is the clincher, it straddles two heroic eras. Tazio Nuvolari raced against it, but Stirling Moss's first F1 race was in 1951 and was won by Fangio in a 158/159! Moss even tested a 159 at Monza when he was a young man making his way in racing – Alfa Romeo was courting him as a driver for 1952, but instead the company withdrew from racing.

The beginning of the Alfa Romeo 158 story

After World War I, motorsport boomed in popularity worldwide. Alfa Romeo had raced more or less from the company's inception in 1910 (as Anonima Lombarda Fabbrica Automobili or A.L.F.A.), but in

1915 prosperous Milanese mining and compressor-engineering magnate Nicola Romeo effectively bought the company, and Alfa Romeo was born officially in 1920, with a remit from its owner to carry on as before, with the aim of making sporting and race-winning cars.

After Vittorio Jano joined from Fiat in 1923 to become Chief Engineer, Alfa started to become the pre-eminent force in both grand prix and sports-car racing. Alfa won both grands prix and endurance events, such as the Le Mans 24 Hours, Targa Florio and Mille Miglia, multiple times. Alfa Romeo became the byword for Italian motorsport and passion. Mussolini owned several Alfas, and all over the world boys grew up wanting an Alfa Romeo, while regarding the team's top drivers as heroes. A young team driver, and later manager, called Enzo Ferrari was learning his trade at Alfa as well...

Scuderia Ferrari quickly became the racing division of Alfa Romeo, and Jano's wonderful 8C 2300 was outstandingly successful on road and track. Other successful cars followed but, as the steamroller that was German-funded domination of GP racing took hold, Alfa began (like everyone else) to fall behind. Jano, Ferrari and others thus hatched an idea for a new car for the smaller single-seater 'Voiturette Formula' for cars powered by engines up to a maximum of 1.5 litres. For this formula, engine design was free, and forced induction could be used. The German teams were not competing in this class, so here was a chance for Alfa Romeo to return to regular winning ways and restore some national, as well as company, pride. It must be said though that Alfa Romeo and Ferrari had high hopes for the new 1937 12C-37 GP car, and they were already looking

▶ Ugo Sivocci pilots his Alfa Romeo RL Super Sport to victory on the 1923 Targa Florio, the first event in which Alfa Romeo's entries featured a green cloverleaf on a white background. This symbol would later become the hallmark of Alfa's competition cars and sportier production models.
(Centro Documentazione Alfa Romeo – Arese)

at designs for the new 3-litre GP formula to come in 1938, so although the popular perception is that the 158 was developed as a way of winning when the larger cars were failing, this is perhaps only partly true, as Alfa truly expected the new GP car to succeed.

There were rumours that the 1.5-litre supercharged class was going to become the premier GP class, and this also may well have helped instigate the 1.5-litre project. Enzo Ferrari (according to Enzo Ferrari!), however, proposed they design a new Voiturette-class car, as whatever the rule makers decided, it was going to be an important class in the future. Ferrari's idea was to design and build it at Modena as 'a racing car of my own'. The plan was greeted enthusiastically by Mussolini, Jano, and his second in command, Gioacchino Colombo – a man he called his right arm. Colombo had in fact been working on just such an idea, making sketches that dated back as far as 1935 and which clearly laid out the specification for the 158. So, whether it was actually his idea, then taken up by Ferrari, or Ferrari that had the idea then realised Colombo had already been planning it, we may never know. Whoever claims credit, Ferrari was setting in motion the design of a car that was so good it took him nearly two years – beginning 12 years later – to build a car that could beat it!

The birth of the 158 – the team assembles at Modena

While the car is, and was always supposed to be, an Alfa Romeo, make no mistake – the 158 was conceived by the Alfa racing team, which meant Ferrari, as by 1937 Scuderia Ferrari was 80 per cent owned by Alfa.

In late April 1937, Colombo was despatched to Modena with his half-finished ideas under his arm, to begin work on the first car to be designed in Ferrari's works – the 158. Enzo Ferrari acted as project manager and progress chaser, running around local suppliers for components and materials, but also assembled a small team under his auspices; Luigi Bazzi (engine, Ferrari's right-hand man and another engineer who had trained under Jano), Angelo Nasi (chassis and front suspension, also on loan from Alfa) and Alberto Massimimo (gearbox, a young engineer Ferrari had hired directly for the project).

By day, this small team worked together in a tiny room in the old Ferrari works in Modena. They also spent their evenings together in the trattoria, and the car took shape quite quickly because it became their whole life. The resulting masterpiece was not exactly revolutionary, but it was absolutely the state-of-the-art of Alfa Romeo design in every aspect, beautifully realised (with a few initial teething troubles) and as such was ahead of all the opposition in both drivability and lap time. It would remain so, with some development, for almost 15 years.

1938 – the Alfetta proves its potential

At the end of 1937, while Colombo's team was working away in Modena designing and building the first 158, which it was planned to race in 1938, Jano's team was working on a car for the new GP Formula in Alfa's Portello works.

▲ One of the world's greatest-ever drivers made his name with Alfa Romeo, Tazio Nuvolari, 'the Flying Mantuan', seen here on the grid for the 1932 Italian Grand Prix at Monza in an Alfa Romeo P3. *(Centro Documentazione Alfa Romeo – Arese)*

◄ Gioacchino Colombo – father of the Alfa Romeo 158. *(Centro Documentazione Alfa Romeo – Arese)*

However, Alfa's manager, Ugo Gobbato, looked at the outsourced racing project 200km away and decided to bring it back in house. This move marked the beginning of the end for the Ferrari/Alfa partnership. The six unfinished 158s (including four finished chassis and one car which was complete enough to have been driven on a local road – yes, they tested race cars by driving them on roads near the factory!) were moved back to Portello, along with jigs, tools and everything else that had been sent to Scuderia Ferrari when they had taken over the operation. Enzo was given the job of managing the newly formed Alfa Corse, a task he never took to and from which he was dismissed by Gobbato a year or so later, albeit with a generous severance package.

Ferrari's severance deal included an undertaking not to take part in motorsport for four years, something that the war made irrelevant anyway. He moved back to Modena, started making small aero engines and machine tools for the military and, in 1943, was able to move to a larger site ten miles south in Maranello. He started making cars in this location after the war under his own name and, in the end, did fairly well I'd say... but that's a different story.

At least one 158 had been completed before the move to Portello but, delayed by the move, the first proper test of the 158 was on 5 May 1938 at Monza – although some vehemently dispute that date. It was apparently trouble-free and successful. The Alfetta's interrupted gestation was over, and within three months it was a works racing car, measuring up against the previously dominant Maseratis in the 1.5-litre supercharged class.

▼ **The three Alfettas, Nos 14, 24 and 26, get underway for the 158's first-ever race, the 1938 Coppa Ciani Junior. This was a Voiturette support race to the main Coppa Ciani, on Sunday 7 August. They won, of course, with Emilio 'Mimi' Villoresi in car 14 leading home Biondetti in 24. Note the original bodywork design; less streamlined and, to my mind, less attractive.**
(Centro Documentazione Alfa Romeo – Arese)

Alfa Romeo 158/159 racing history

First race – Coppa Ciani Junior

By the time of the Coppa Ciani Junior, in August 1938, six complete 158s had been built. Three were entered for the race, and they caused a sensation, making every car look old fashioned, tall and most of all slow! They dominated qualifying, and came first and second, with Emilio Villoresi winning from Biondetta.

The Alfas retired from their next race, the Coppa Acerabo, only one week later, and were not even as quick as the Maseratis. What had happened in a week? The circuit ran from the seafront into the high mountains and the carburation could not cope with this. Alfa had been humiliated and learned a lesson to guard against complacency.

Alfettas came first and second in the Voiturette feature race at the Italian GP, Monza, in September 1938. A week later, Alfa entered four 158s in the Circuito Di Modena Voiturette race. They were quickest in practice, but all retired with engine-bearing issues.

Alfa had some work to do over the winter, as they'd designed the best Voiturette racing car, but only when it held together.

1939 season

The first race of the season was the Grand Prix of Tripoli, on 7 May 1939, in Libya. Unfortunately, the Alfas all retired, in a frankly shambolic way.

The race was run to Voiturette rules, and Maserati surprised all by entering a streamlined 4CL, which

was the fastest car in practice but lasted only one lap in the race. However, the biggest surprise was that Mercedes had produced a Voiturette – the W165. It had been built and tested in secret very quickly. The streamlined 4CL of Luigi Villoresi was the fastest in practice, and the two W165s of Hermann Lang and Rudolf Caracciola were close behind.

Alfa sent six cars. They qualified behind both the Merc and the Maserati streamliner, but all hit problems in the race and retired. Costantini, nervous of the heat, had instructed the mechanics to reduce the cooling-system pressure to prevent overheating. This, of course, actually caused the overheating, because the system was designed to pressurise!

The Mercedes 1–2 had proved a point, and the W165 never raced again.

The 158s won four of the next five races they entered, including the 1940 Tripoli race, although with no German or British cars there was no real opposition. Then war came.

The first post-war race

The first race to feature the 158s post-war was a humiliating defeat for Alfa, on 9 June 1946, on the tight twisty St Cloud circuit near Paris. Both cars, driven by Emilio Giuseppe 'Nino' Farina and Jean-Pierre Wimille, retired with failed clutches. They were quick though, and Alfa could see its by now nine-year-old car design was still the class of the Voiturette field.

The famous winning streak starts

The next race, on 21 July 1946, was the GP of Nations in Geneva. A team of four Alfas

was entered, two of which featured the new two-stage supercharging, which gave around 254bhp@7,500rpm. The two heats were both won by Alfas, and the final was an Alfa-dominated event too, with Farina leading home Trossi and Wimille. This race started the Alfetta's winning streak, which lasted until 1951.

Formula 1 begins

For 1947, the FIA created a premier GP class, then called Formula A. This was stated as 1,500cc supercharged cars (the old Voiturette class in which the Alfetta had run previously), or up to 4.5-litre

▲ **Rudolf Caracciola on his way to second place in the Tripoli Grand Prix, 7 May 1939, in the Mercedes-Benz W165.** *(Daimler AG)*

▼ **Consalvo Sanesi (24) and Count Carlo Trossi (30) in their 158s at the 1947 Italian GP.** *(Centro Documentazione Alfa Romeo – Arese)*

naturally aspirated cars. Thus, for 1947, Alfa's baby racing car suddenly became their main grand prix car, when it was already eight years old.

The first race run to the new rules took place on 1 September 1946 – the Grand Premio del Valentino – and was won by Achille Varzi, in an Alfetta of course, with his team-mate Jean-Pierre Wimille second.

Alfa chose to race in only four events in both 1947 and 1948, all of which they won! Alfa then chose to retire completely from racing in 1949 and the world thought the Alfettas may never race again. The point had been proven and the legend had been created. These were the fastest cars of their type, both pre- and post-war. Alfa as a company had also benefited from the Marshall Plan to rebuild Europe, so racing seemed publicly inappropriate whilst receiving 'aid'. They also needed to concentrate on finishing and launching the new mass-market 1900 saloon, which came out in 1950. It was Alfa's first production-line, monocoque, high-volume car, and a real change of direction for the company.

The triumphant return

Engineers within Alfa continued to develop the 158 though, finding yet more power from the engines, just in case the cars were needed. The planets then aligned in 1950 for a return, which by rights, with a 12-year-old car, should have been a failure...

Firstly the conditions surrounding the Marshall Plan were relaxed. Secondly those advocating the need for engineers to work on the 1900 project rather than racing were now saying; 'We need to race to sell the 1900!' Thirdly, the FIA announced that the Formula A (now Formula 1) that Alfa Romeo had dominated with the 158 since 1938 would now be a proper World Drivers' Championship, the first one in the history of the sport.

Alfa hired Farina back to drive for them, plus a new driver who had started to make waves in Europe, called Juan Manuel Fangio. The experienced veteran Luigi Fagioli completed the team – 'The Three Fs', as they would become known. The cars were modified further, with 350bhp now being achieved, improved brakes and a stronger gearbox.

Alfa entered every round of the seven-round F1 Championship, except for the Indy 500, which all the European teams ignored.

There were a number of races run to F1 rules before the first championship event on 13 May. Alfa entered a single car for the San Remo GP on 16 April, initially for Farina, but when he broke his collarbone in an F2 race, new boy Fangio managed to persuade them they had nothing to lose byputting him in the car, saying, "I'm an unknown and if I lose, Fangio loses. If I win, Alfa Romeo wins." Against no fewer than seven Ferraris and eleven Maseratis, Fangio drove a measured race in the damp conditions and won by over a minute from Villoresi's Ferrari 125.

The start of modern-day F1 – Silverstone 1950

The British Grand Prix took place on 13 May 1950, and Ferrari elected not to go to Silverstone, in order to concentrate on car development. Alfa brought four cars, one each for 'The Three Fs', and one for English driver Reg Parnell.

Farina was on pole, and both of the other 'Fs' posted identical times, for second and third on the grid, with Parnell fourth.

The Alfas simply drove away from the opposition in formation, but Fangio hit a hay bale (1950s spectator protection!) and eventually retired. Farina won from Fagioli and Parnell.

Ferrari was back for race two, in Monaco. Farina led away, but was soon overtaken by Fangio. Farina spun at a damp Tabac, and a huge pile-up developed, which eliminated nine cars. Fangio stopped for fuel only once, whereas second-place Alberto Ascari's Ferrari 125 needed two stops, so Fangio cruised to victory by a lap.

That left Farina and Fangio jointly leading the championship, but Farina won the next round, the Swiss GP at Bremgarten. Fangio then won at Spa, significant because on the long Masta Straight, the new electric timing equipment timed Fagioli at 200.75mph, which is just awesome for a 1.5-litre car with no safety equipment running on skinny tyres – my admiration for these men knows no bounds. I think this was the first time an F1 car was timed at over 200mph.

At the French GP at Reims, on 2 July, Fangio, driving an experimental 158 with even more boost pressure, simply demolished all, including his

▼ **Alfa Romeo entered the new F1 World Championship with a car that was unquestionably the best available, despite being a 12-year-old design. They also had the most professional and best-equipped team, as this lovely shot from the paddock of Silverstone in 1950 shows.**
(Guy Griffiths Collection)

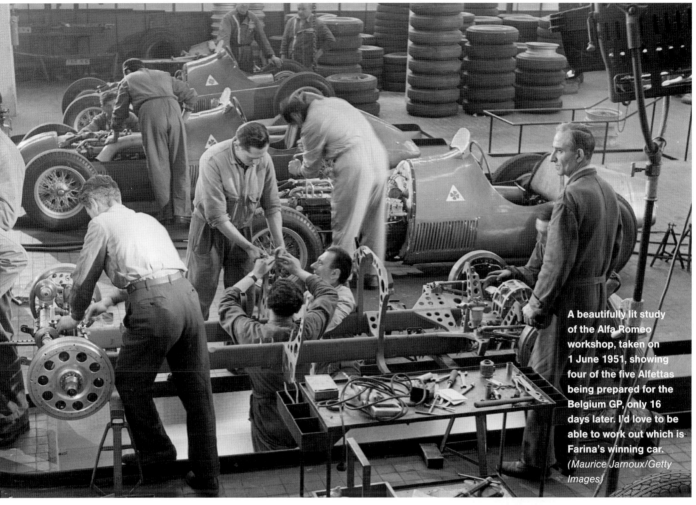

A beautifully lit study of the Alfa Romeo workshop, taken on 1 June 1951, showing four of the five Alfettas being prepared for the Belgium GP, only 16 days later. I'd love to be able to work out which is Farina's winning car. *(Maurice Jarnoux/Getty Images)*

▼ ▶ Farina crosses the line in his 158 to make history and win the first ever F1 World Drivers' Championship race, at Silverstone on 13 May 1950, and celebrates afterwards. *(Centro Documentazione Alfa Romeo - Arese)*

▲ Luigi Fagioli leads Juan Manuel Fangio during the French GP at Reims on 2 July, 1950. *(AFP via Getty Images)*

▶ Mechanics unload the Alfettas prior to a test session at Monza on 1 June, 1951. *(Maurice Jarnoux, Getty Images)*

▼ Fangio splashes to victory in the Swiss GP of 1951 in a 159-specification Alfetta. *(Centro Documentazione Alfa Romeo - Arese)*

team-mates, with the sheer speed of this 370bhp machine. The upstart and relatively unknown Argentinean had become the surprise star.

Alfa also entered a small number of non-championship races in 1950, all of which they won.

All eyes were on the Italian GP at Monza on 3 September. It proved, that it's not over until it's over! Alfa entered five cars, 'The Three Fs' plus Senesi and Taruffi, and Ferrari finally debuted their new 4.5-litre 375 V12 for Alberto Ascari. Fangio was quickest in practice, ahead of Ascari and Farina. In the race, Fangio retired with a broken gearbox and Ascari had mechanical dramas (although he finished second by taking over his team-mate Serafini's car), so Farina emerged the clear winner by over a minute, and thus clinched the championship.

1951 season

The naturally aspirated Ferrari 4.5-litre V12 375 F1 was a car that Enzo Ferrari effectively dictated would be built after he'd failed for three years to get on terms with the car he had overseen the design of 12 years earlier! The 375 was not convincingly faster than the Alfa, at least not until late 1951, but it was more economical than the now crazily thirsty Alfa.

The Alfas had struggled in torrential rain in a non-championship season-opening event at Silverstone, but the championship was Alfa's to lose, and they continued where they had left off, with Fangio winning the first race in Switzerland.

Farina took a remarkable victory on the twisting road circuit of Dundrod in the non-championship Ulster Trophy, and followed it up with a win in the Belgian GP at Spa, driving a swing-axle 159. Fangio was back to winning ways at Reims, France, on 1 July, taking over Fagioli's car after the magneto had failed on his own, and thus sharing the points.

The next race, at Silverstone, was the beginning

of the end for the Alfetta. Now equipped with even more power, the Alfettas still didn't manage to set the fastest lap in qualifying, being beaten by José Froilán González in a Ferrari 375. This was González's day of days and Fangio always said the Alfetta didn't lose, but rather that his fellow Argentinean, González, won. The cracks in the now 13-year-old Alfetta were starting to show. Extra power had been found by running high boost and using fuel that was 98 per cent methanol, but at the cost of prodigious thirst; Fangio used over 200 gallons of fuel in that one race. The naturally aspirated Ferrari was positively economical in comparison, which meant that it was both lighter and spent less time in the pits.

At the next championship round, at Nürburgring in Germany, Fangio damaged his car in practice, and the circuit's notorious bumps were making the chassis flex so much that Alfa actually built a stiffer frame and rushed it to Germany after practice. Farina retired on lap 8, the chassis flex damaging pipes, which created overheating. Fangio's gearbox gradually lost gears and he finished third. Alberto Ascari was victorious in the Ferrari 375.

The Italian Grand Prix at Monza on 16 September saw Fangio grab pole position. The Alfa's engine now produced 425bhp@9,300rpm, and the cars of Fangio and Farina were the two fastest qualifiers. In the race, Ascari (in the Ferrari) and Fangio battled, but Farina retired on lap 6 with engine problems, and Fangio also retired on lap 39 with a broken piston. Farina took over Bonetto's car and achieved fastest lap on his way to a shared third place. Alfa appeared to have discovered the Alfetta's limits, and after almost 14 years of constant development the gallant little car looked vulnerable. However, Alfa insiders insisted to their death that their cars had been tampered with that day. The inference was that a certain ex-employee had paid former colleagues to nobble Alfa, though nothing was ever proven. You can become bitter about your first love...

Farina took two easy non-championship victories at Goodwood, but the final race of the season, at Barcelona in Spain in October, was a straight head-to-head for the championship between Fangio in the Alfa and Ascari in the Ferrari. Ascari was on pole from Fangio, but Ferrari chose to use 16in wheels and this proved their undoing, as the tyres started to break up on the rough track surface and Fangio, concentrating on driving smoothly to avoid having the same thing happen to his larger tyres, was able to take the win. Hence, Fangio was the World Champion.

Alfa had decided to retire at the end of 1951 if they won the championship. The Alfetta was 261kg heavier than it had been in 1938 after all the modifications, and although tests had shown that 450bhp was possible, handling, fuel consumption and tyre wear were suffering. Alfa had received a government grant of 100 million lire to race in

1951, but Colombo calculated the budget needed to remain competitive was five times that. So, eventually, the same offer for 1952 was turned down, and Alfa Corse activities were reined back to sports and saloon car racing.

The Alfetta had regained Alfa's pride by winning its last race, after a frontline competition life that had started in 1938. No other car has won all the races it entered in a season – the Alfetta did it several times! It won its first race as a Voiturette and its last race as a full-on GP car. Between 1947 and its retirement in 1951, it won 31 races out of the 35 it entered!

If that's not awesome I don't know what is.

Sadly, some Alfettas were broken up, but all nine engines survive and seven cars still exist. Alfa Romeo owns five of those, and one is demonstrated regularly by trusted individuals such as Alfa's 2020 F1 star Kimi Räikkönen.

▲ **Fangio pulls in driving my favourite No 22 to claim victory in the 1951 Spanish Grand Prix, the final race of the year, on 28 October. The win gave him his first F1 World Championship title and Alfa Romeo their last.**
(Motorsport Images)

▼ **Alfa's current Formula 1 driver, Kimi Räikkönen, demonstrating an Alfetta at Silverstone in 2019.**
(Centro Documentazione Alfa Romeo - Arese)

CHAPTER 4

Plan, plan, plan

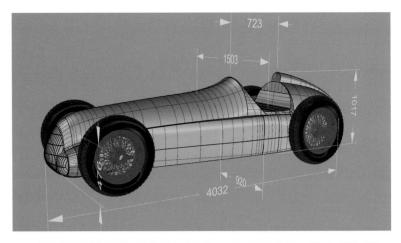

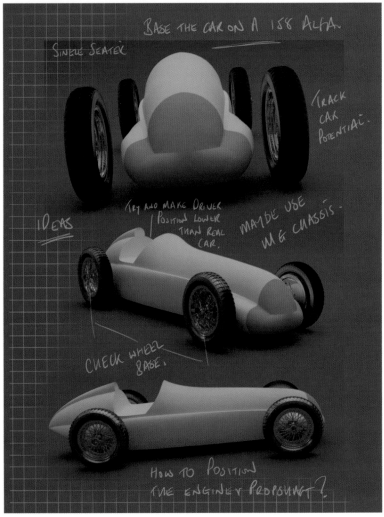

At the outset, building a special may seem like a daunting proposition, but there is a lot of truth in what Henry Ford – the man who originally brought motoring to the masses – said: "Nothing is particularly hard if you divide it into small jobs."

For me, the true art of building a special is allowing the car to take shape organically. Being flexible throughout the build is essential. It's also more fun and liberating that way. If you want to make a change, just make it. Unlike restoring an existing car, you are not bound by the predetermined design and engineering of that specific model. That being said, to start without a plan is foolish. So, the first stage in my planning was to turn to my pal Paul Cameron and ask him to produce some accurate dimensional CAD (Computer-Aided Design) drawings.

The use of modern-day CAD makes car building easier – it is very accessible, and it's also simple to source a local firm to assist. For this build, I was not building a recreation, more of an 'homage'. A great deal of thought went into making my car exactly as I wanted it – after all, this was a car for me. Using CAD drawings enabled me to engineer my car with minute accuracy. If you decide to build something that looks different, then my advice is to draw it (even if only by hand), print it and place it on the wall as inspiration (and motivation) throughout the build. It's always nice to glance over at a picture of what you hope to create.

When settling on an initial design, it may be that you place pictures of a number of different cars on the wall, and choose the best elements from each vehicle. When I designed my take on a classic 1960 sports car – 'The Comet' – I drew inspiration from Aston Martin and Ferrari at the front, Lotus in the middle and the Ferrari 250 GT SWB and Maserati at the rear. It was a mash-up of my favourite '60s cars, yet its overall shape is unlike any of them.

In simpler times, and before I evolved to CAD, I used a pen and paper. It's a great way to get started. One of the concept cars I built started with a drawing, and I then built a buck and mould, and the fully finished, road-legal car was invited to appear on the Cartier 'Style et Luxe' Concours d'Elegance lawn at the Goodwood Festival of Speed. What started out as a simple drawing on a blank page evolved

◀ **To start without a plan is foolish. These tremendous 3D renderings, from my good friend Paul, really helped me to visualise the body.** *(Paul Cameron)*

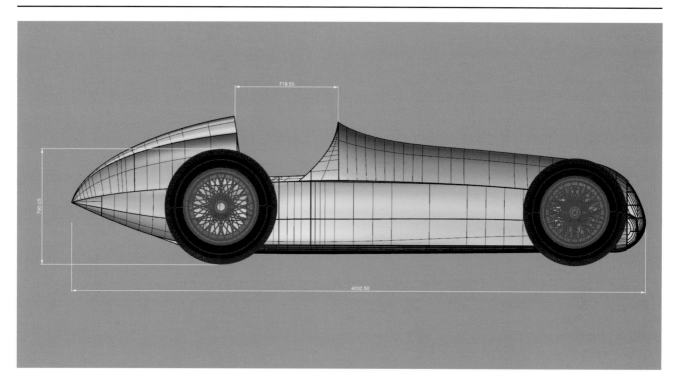

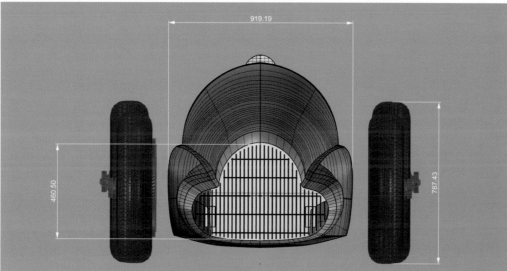

▲◄ You don't have to use professional CAD drawings, like these, for your build, but if you know anyone who can operate a CAD system, you might want to call in some favours.
(Paul Cameron)

▼ It's always mind-blowing to see a car for real, after it has only existed 'on paper' for a long time. This is 'The Comet' – a mash-up of my favourite '60s cars.
(Darren Collins)

▶ Another great designer pal, Dave Clark, sketched this image of a Barchetta-inspired build. Four of these were made. *(Dave Clark)*

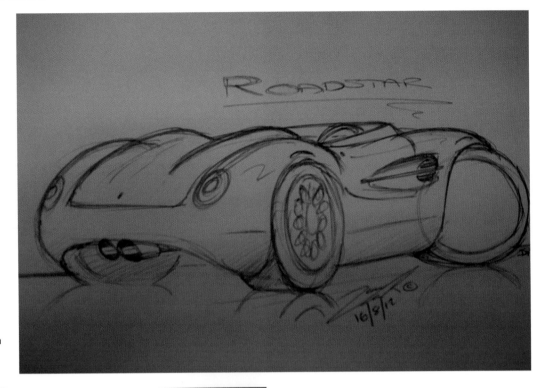

▼ A body panel just 'born' from the mould for the Barchetta build, inspired by Lotus, Aston and Ferrari. *(Ant)*

into a road-legal reality that I'm still incredibly proud of.

That is exactly the method I have used for designing this special. I started with the idea – the bare basic concept. First, I got a hold of articles, books, pictures and posters of the Alfa 158, and of course there is also a wealth of information online. I studied the history and heritage of the car and thought about what drew me to it when I was a kid. I always kept these items of information to hand throughout the build, so that I could refer back to the drawings and pictures for inspiration, or simply stare at them during the hundreds of tea breaks!

Tools and working space

At the very least it will be necessary to invest in a decent range of hand tools, an angle grinder and a welder, and to ensure sufficient protective and safety equipment is available. Enough space needs to be created to build the car freely, whether it's in a garage, a shed or on a driveway. I was lucky, as I had access to the *Wheeler Dealers* workshop during a break between productions, and double lucky it's pretty well kitted out, with way more than I need for a special build!

For some of the 'Alfa' build, I left the confines of the workshop and turned to various small businesses to assist me. There are so many great companies out there, all ready and willing to help

◀ Me proudly standing next to the completed car on the Cartier Concours lawn at The Goodwood Festival of Speed. *(Julius Thurgood)*

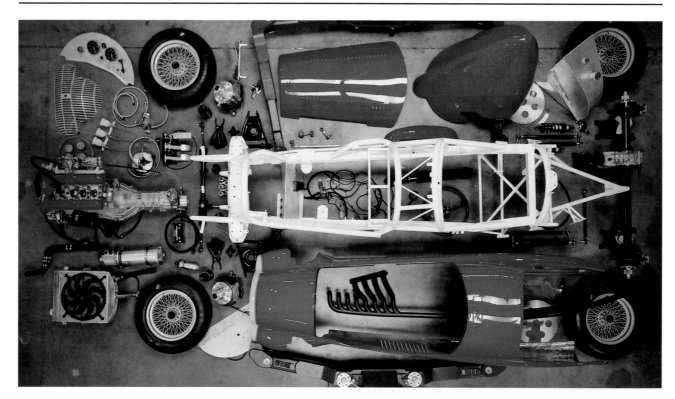

your build. When it comes to items like driveshafts, propshaft and paint spraying, it's worth investing in external help. Be armed in advance, ask friends for referrals, or look online to find and plan these services locally. The DIY car sector is really well supported and connected, and there is a whole support network of car clubs, forums and car fans – it's worth making use of these amazing people.

Parts

When you break it down (into a series of small steps), this build is actually relatively simple, with very few components. It's surprising when you conduct an inventory of parts how few parts are used. In the following pages I will explain step-by-step the details of the parts I used to build my car, but feel free to decide on your own requirements and make bold changes to suit your own build, the way you want it.

In this book, I have not focused on the mechanical details of the donor parts. The strip-down, renovation (if necessary) and rebuild of components are best served by other publications (such as Haynes Manuals!), which tackle those specific jobs in depth. I have provided details of the parts I used, and you may choose alternative parts for your build. This is simply my story of how to build a special from the specific parts I chose.

Skills

Do not feel that you have to limit your special build due to the level of your practical skills, or worse still, your own perception of the skills you have. It's OK to try

and to fail, or simply to ask for help. Some of the skills required to finish this car can be challenging. There is a lot of welding and machining for example, and if that is out of your comfort zone, why not go and get some lessons, and add that new skillset to your repertoire? Your skills and experience will grow alongside your car, making this build even more rewarding.

Let's make a start!

With the planning done and the appetite whetted, it's time to start. For me, there is real energy to be gained from action! My dream is that a whole host of home builders will be inspired by this book to create their own unique cars. You have to make a decision to start your build, then roll up your sleeves and just get stuck in!

There endeth the back-story, and now for action. It's time to start building...

▲ **The finished car with everything powder coated, or polished, ready for assembly. This is a magical moment.** *(Mike Brewer)*

▼ **Many people came together and chipped in to help me to complete my build. Don't be afraid to ask for help.** *(Chris Hill)*

CHAPTER 5

Chassis

Like most specials I have built in the past, I opted for a classic ladder chassis. From an engineering perspective it's really simple, and the perfect platform on which to evolve and develop a car. The chassis alterations I will show later eventually transform it to more of a hybrid space-frame chassis. For me, building a special is in some ways similar to building a house, working from the foundations up. The chassis effectively provides the foundations.

For my car, I avoided building a brand-new bespoke chassis from scratch (I have half an eye on road-registering my car in California later). Getting hold of a rolling chassis or a part-assembled project is actually quite easy, as there are so many stalled builds out there. I set about finding my perfect starting point, and I had a ton of choices. Online car sites are littered with 'rolling chassis' projects for sale. I wanted my car to have European heritage, so I settled on a 1952 MG TD I found online. I bought the car blind (which I usually frown upon) and shipped it from the other side of the US. You can see in the photographs the condition in which she arrived at the workshop one afternoon. As I've already mentioned, my first-ever car was a 1970s MG Midget, so I admit that building a special on a British chassis from the octagonal stable was heart-warming on an emotional level.

She looked like a car beyond repair, and as an MG TD this would no longer have been a sensible or viable restoration. However, as the donor base for my special, she was absolutely perfect. I really love the fact that this deceased MG is getting a second shot at life. What attracted me to this particular car was the fact that the rolling chassis part of the car had clearly had some restoration work done already. It was rust-free, painted, and complete.

In Europe, an MG TD may prove to be a tricky and possibly pricey donor, as around 75 per cent of MG TDs made were exported to the US. Getting

◄ **The donor MG TD was someone else's stalled restoration project.** *(Chris Hill)*

▼ **Sometimes the work required isn't worth it, but for my project this was the perfect base.** *(Chris Hill)*

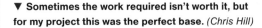

THE CHASSIS – A CAR'S FOUNDATIONS

A ladder chassis is one of the foundations of car making, which after all, grew from the carriage trade. Since the 19th century, carriage makers had taken two stout wooden rails, joined them together at each end and built a body on that base. That principle continued to work when a small engine was added and the rails were made from metal. The chassis needed to be narrower than the car in order to minimise the flex in the structure, but built correctly a ladder chassis was very strong. The Ford Model T commercialised that method of building cars, and others followed that lead. The American market's commercial imperative to produce revised bodywork for each model year encouraged that method of production, as the chassis could remain basically the same, but could receive a new body with bigger fins or extra portholes.

However, once metal-pressing technology improved and production volumes increased, the monocoque or 'unibody' gradually became the preferred configuration, leaving ladder chassis to be used on sports cars, commercials and 4x4s.

▲ Ladder chassis can be very strong, as this wonderful shot of a Model T working for its living shows. *(Ford Motor Company)*

▼ Steel-construction pioneers: the Lancia Lambda of 1922 (left) and the 1934 Citroën Traction Avant (right). *(Lancia Automobiles/Citroen, Fonds de Dotation Peugeot pour la Mémoire de l'Histoire Industrielle)*

hold of a TD here, in America, was really simple. In fact, I was spoiled for choice. So, choose a donor that's accessible based on where you're located, and remember, it's fine to look beyond MGs. There are many other ladder-chassis cars available, you've just got to look for them.

I would also need a second donor car, for the running gear, but we'll come to that later on.

The strip-down of the donor chassis

The aim of the game here was to simply remove everything to leave a bare, naked chassis. I found the MG TD body came away really easily. It's usually held on to the chassis rails by a few bolts, but mine was actually held on by cable ties for transport, so a couple of snips and it was away! For strip-down, using any tool is fair game, no rules, just don't rush. At this stage, you want to preserve everything that's part of the actual chassis – all the brackets and threaded holes – as you never know what you might need later. Also, it's a good idea to use a camera to document the dismantling, as it may help when refitting parts

▼ The ladder chassis from an MG TD is perfect for a build like this, and in the US there are plenty available to choose from. *(Chris Hill)*

▲ **Rolling the donor into the workshop on the first day. The body came off in about five minutes.** *(Master Mechanic)*

▲ **The child in me was always going to sit in the car and assume 'the position'.** *(Master Mechanic)*

later, but better than that, it will enable you to share the details of your build in years to come.

As I was stripping down, I made sure I looked closely at the chassis condition, watching out for damage, corrosion and pitting. I was lucky that my donor chassis was really solid and required no patch repairs.

Starting with the main upper body, you can try and lift it all away in one go, or strip it piece by piece. I chose to remove the whole lump, as I didn't need any of the MG body for my special, so why waste time taking every nut and bolt off? Plus, my wooden-framed body could be useful to an MG TD restorer, who will need all those small, hard-to-find parts. Using a two-post ramp made light work of lifting the body off, but equally a couple of strong helpers could lift it away by hand.

With the body put to one side, I supported the chassis on my two-post ramp, but four equal-sized axle stands would do just as well. As an extra precaution, I also clamped my chassis to the ramp to aid stability, as the weight comes down really quickly once you start removing parts. I only needed basic hand tools to dismantle the chassis components, first removing the wheels, as these were going in exchange for wire wheels later.

I had no engine or drivetrain to deal with, but if you do, now is the time to remove it. Again, no rules, anything goes, provided you don't want to reuse it of course. It may be that you choose to retain the MG TD engine, but at only 54bhp, I was holding out for something a little more powerful!

If you don't want to use the engine that came with the donor, it may be worth posting it for sale online. It might be just what a builder down the road (or halfway across the world) is looking for, and before long your build could be funding itself, as you sell off the parts you don't need.

I then turned to the front end and removed the steering rack, which was easy – four bolts and two rod-ends to disconnect. At this early stage I decided to retain the complete front end. One of the reasons was that the MG TD uses rack-and-pinion steering – the first-ever MG to do so. It was also the first to be fitted with double-wishbone front suspension, and to be offered in left-hand-drive – all great news for me. The earlier MG TC was fitted with an old-style Bishop Cam steering box and drag-link steering, with a beam front axle – which would not provide the handling ethos I'm looking for on my race-inspired special. The responsive feel from a rack-and-pinion system makes the steering so much more direct, and that's why pretty much every modern-day car still uses rack-and-pinion steering.

◄ **The tyres on the MG TD wheels were rock hard. They are still kicking around if anyone wants them.** *(Master Mechanic)*

▲ Strip any components from your chassis that you don't want to be shot blasted, which will be most of them. *(Master Mechanic)*

▲ Everyone on the TV-show team helped. That's Scott, the cameraman, helping me carry the chassis. *(Master Mechanic)*

I was a tad more careful when removing the front suspension parts, as I knew I wanted to retain them. It often helps to take a set of photos as the parts are dismantled, and to keep a log, to help with the rebuild later. Safety is always paramount, and so care has to be taken when removing the suspension springs – a set of suitable purpose-made spring compressors must always be used, which compress the springs to allow them to be removed safely from the car. The spring tension can then be released by safely loosening the compressors on the bench or floor.

Instead of removing every component individually, I took a far simpler route and dropped the whole front suspension assembly off, one corner at a time. It was day one in the workshop for me, and I wanted to end up with a naked chassis – I could dismantle the suspension assembly on the workbench later.

My next move was to remove the live-rear-axle assembly that sits on two huge leaf springs. For sure this rear end was going! I needed a more compact unit, capable of dealing with a touch more power (we will come to that later). So again, I dropped the assembly off the chassis in one lump – I didn't even remove the springs from the axle – and posted it for sale online, for a needy MG TD restorer to pick up.

With just a handful of small items left to remove, such as check straps, brake-pipe clips and bump stops, I had stripped my chassis in less than an hour. A simple start, and a great time to pause, stand back and admire the bare chassis that was left. It may appear a little basic – just a naked ladder chassis – but from here, great things would grow...

Checking the chassis for damage and distortion

It goes without saying, that if possible, the chassis should be checked for damage and distortion before you purchase it, but it is still worth checking again at this stage, and is easier once all the components have been dismantled.

Starting with a straight chassis is important, and if

▲ Nice rear end! This whole rear axle and suspension assembly was sold on to an MG restorer. *(Chris Hill)*

◄ The fully stripped chassis before heading for shot-blasting. I had my fingers crossed it wasn't full of filler. *(Chris Hill)*

▲ **I could have taken the stripped chassis to a specialist to check it was square, but I was happy with a tape measure. Thanks, Christina.**
(Master Mechanic)

the donor doesn't come to the party in good shape, then there is no point using it. When everything is stripped off the donor chassis, it's little more than an intricate ladder, and it has to be right. In the past, I've built cars by starting with my own scratch-built ladder chassis, because it was easier than straightening a bent one, or finding the exact chassis I wanted.

It's worth taking time at this stage to closely study the condition of the chassis, checking carefully for odd-looking creases, old repairs, holes, or thin metal. If there is any suspicion that filler has been used to hide old secrets, a magnet can be used to check – a magnet will not stick to filler!

If the workshop floor is perfectly flat, the chassis can be placed on axle stands, and measurements can then be made from the floor to the lower edges of the chassis at various points. The measurements on each side should be the same. Try to get a feel for how it should be.

The ultimate check would be to make friends with a body shop and take the chassis there. Body shops have alignment jigs and can tell you to the millimetre how straight the frame is. They will attach the chassis to the jig, and using a series of tools, they can measure to specific points on the frame, checking they match. I was comfortable taking these measurements myself, using a tape measure and cross-referencing with the dimensions provided on the original MG TD chassis drawings I got hold of.

Cleaning her up

Sand blasting, shot blasting and bead blasting are all slightly different techniques, but essentially it's the same process of firing a very hard granular material out of a gun, with the intention of removing paint or corrosion. The word 'blasting' kind of gives it away!

I used garnet as my blasting media. Garnet is actually a crystal, crushed into grains that look like sand. Garnet is great because the crystals maintain really sharp edges, even when very finely granulated. It's those sharp edges hitting the steel at high speed that cuts away the paint. Although noisy and dusty, it's incredibly quick. There is a whole industry associated with garnet mining, for both industrial use and, less frequently these days, for gemstones.

This cleaning-up process could be tackled by hand, using a wire brush and some abrasive paper, but my advice is to go and make friends with the local blasting specialist. They are often noisy, industrial places, and it might not be cheap, but it's worth it, as you really get to see the state of the metal underneath. Tackling this cleaning process by hand is doable, but will take simply ages.

Before blasting, I took some time at this stage to protect the front shock-absorber mounts. On the MG TD these are threaded into the chassis. I placed some rubber bungs in the four holes on each side to protect the threads during blasting, as I would need to use these threads later.

If you are going to carry out blasting yourself, it is vital to protect yourself with the correct equipment. I did the blasting myself at a professional facility that supplied the right gear.

When the chassis was stripped back to bare metal, I took the time to check over the entire structure for any sins that the blasting process may have uncovered. Paint is very good at hiding rust and filler. It turned out that, as my gut feeling had told me, my 1952 metal was in really good shape – Old Father Time had been kind, and I had no patching or weld repairs to do at all.

If any holes are revealed, it will be necessary to investigate the extent of the damage. The best approach is to carefully cut away the bad material with a small angle grinder, and keep going until good metal is found. The key is not to be shy here, and not to pretend it's in better shape than it is. If there's any doubt, ask somebody suitably qualified for a second opinion.

Small repairs can be made by cutting steel patches to match what has been cut away. The patches need to be welded in, then the weld ground back, and the process repeated until the repair looks like the original chassis. It's essential to start with a strong base for the chassis, and to maintain standards as the build progresses.

The CAD plan

The next bit was fun, and involved working like a proper car designer. I had the original MG TD chassis diagram from the back of the owner's manual blown up to A1 size at a local photocopy place, then I traced that image onto a new piece of paper to represent what I actually wanted to work with on the workshop floor. It was only necessary to draw the basic chassis, leaving out anything that

▲ Shot/sand/bead blasting – it's all pretty much the same thing. It's hot, loud and dirty. I love it. *(Master Mechanic)*

▼ I was now feeling like a proper car designer, and was starting to remove any of the MG TD chassis that I didn't need – on paper at first. *(Master Mechanic)*

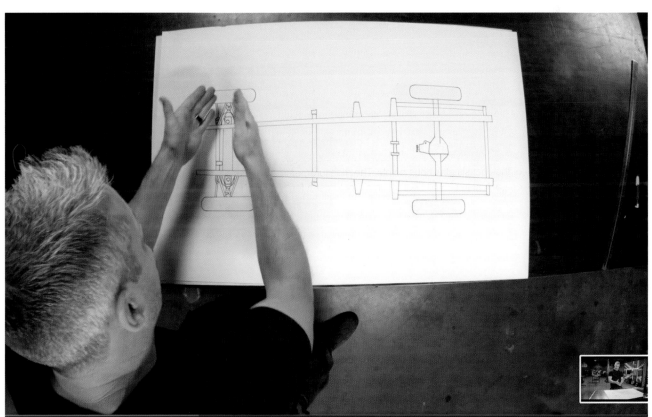

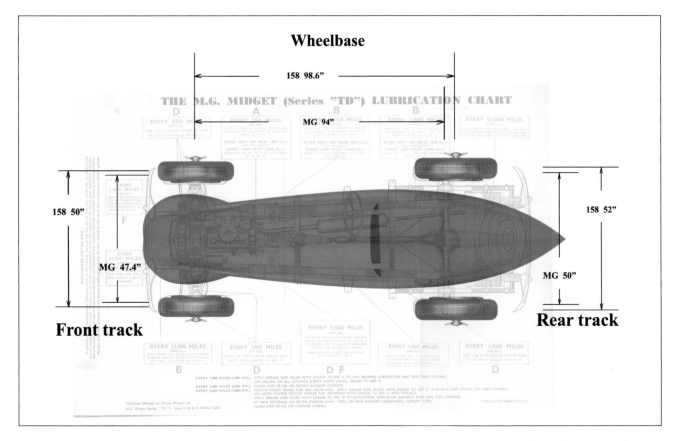

Wheelbase

158 98.6"

MG 94"

Front track   Rear track

158 50"   MG 47.4"   158 52"   MG 50"

▲ **Overlaying the MG TD image with the Alfa body shape was a wonderful moment. Anything not within the body line had to go.** *(Paul Cameron)*

▼ **This cordless reciprocating saw made short work of the chassis metal. Gloves are essential.** *(Chris Hill)*

wasn't needed – this left a simple chassis outline on paper, ready for business.

My friend Paul then created the CAD model of the Alfa body for me, and I was able to blow that up to the same scale as the MG chassis. I did this knowing the wheelbase of the MG TD and the wheelbase of the original Alfa Romeo 158.

The MG TD wheelbase is 2,388mm (94in) and the Alfa 158 is 2,504mm (98.6in). Printing the 158 drawing on transparent paper and overlaying the two drawings was a rather revealing moment. Basically, anything on the MG chassis that stuck

out beyond the outline of the Alfa body needed to go.

I wanted to keep the boat-tailed shape of the 158, and there was no way I was going to compromise the rear of my car with MG TD chassis outriggers and extraneous frame parts poking out at the back. As mentioned previously, that means the cumbersome leaf springs and their mounts are goners too!

▼ **A cordless angle grinder fitted with a 4.5in cutting disc was perfect for removing the unwanted parts from the chassis.** *(Chris Hill)*

It's vital to make a plan prior to cutting anything! It may be that you are choosing to build a two-seater special, or simply using a wider body and will make no cuts. Either way, it pays to be armed with a plan – first in your head, then get it down on paper.

Cutting the chassis

I decided the best tool for cutting the chassis was a 4.5in, hand-held angle-grinder, fitted with a thin (1mm) cutting disc. Once again, safety is the first consideration, so always wear goggles and gloves at the very least. Not only are the flying sparks a risk, these discs are fragile and have been known to shatter.

The disc only got me so far through the outriggers, but they did so really accurately. When I had cut as far as I could go with the angle-grinder, I switched to a reciprocating saw with a metal-cutting blade, to finish the hard-to-reach bits. Again, gloves and goggles are essential.

While I was at it, I also cut off the large bulkhead roll hoop, and the inner crossmember below it, plus the mounts for the pedal box.

I was left with a pile of metal parts that I didn't need, so I put them to one side, in case they came in handy sometime later.

At this point, I pre-marked the longer Alfa 158 wheelbase on the chassis using a permanent marker. I wanted to position the special's front axle in the exact same (lateral) position on the chassis as the original MG TD front axle, so I took a measurement from the centre of the front axle line on the chassis (a line across the front of the chassis between the centre points of the plates with the four threaded holes that the shock absorbers bolt to). I measured 98.6in back along the chassis from this line, and drew a second line across both chassis rails to mark the position where the centreline of the new rear axle would eventually sit. Ultimately, everything from close behind this line rearwards would be cut off the chassis, but not just yet.

At this stage, I could visualise, and get my head around not only the longer wheelbase, but also the increased front and rear track – the Alfa 158 is, after all, bigger than the rather tiny MG TD.

Now I could reflect on the fruit of all that labour. Note I left the chassis rear internal crossmember in place at this stage, as this was roughly where I would eventually sit in my special. I left it in place (I later removed it, once the chassis had been reinforced) to retain some strength in the chassis, keeping the chassis rails rigid and secure for when I eventually came to chop off the rear end.

This was the first of a number of stages working on the chassis, but before making any more chassis progress, the next stage in the build process was to fix the positions of the four wheels. So, leaving the chassis on the ramp, I moved on to the front suspension.

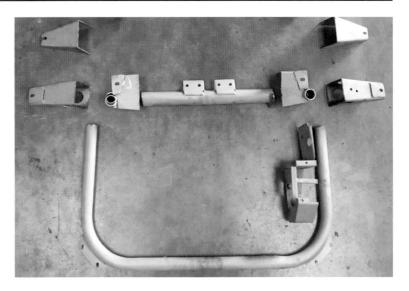

▲ The parts that came off the chassis went into a pile in the corner – they might still be there. *(Ant)*

▲ Marking the rear-axle position on the chassis. *(Master Mechanic)*

▼ Who says you need a load of fancy tools to be a car builder? *(Master Mechanic)*

CHAPTER 6

Front suspension

I kept the MG TD front suspension because it looks great (bear in mind a lot of it is on display), and it's period. The original Alfa 158 used an independent front suspension system with trailing arms and low-mounted transverse leaf springs, plus a combination of friction dampers and direct-acting hydraulic telescopic dampers. There are, of course, better modern suspension solutions out there, but for me, the aesthetics of my final car were very important. It had to look like a period racing car. Newer style 'coil-over-shock' suspension units would improve the car's handling, and losing the brass-coloured trunnions/kingpins, would, overall, be better engineering, but they wouldn't look anywhere near as cool as the existing 1950s units. The old-fashioned dampers, wishbones and uprights were all there, working, on the MG, and I've kept them.

Shock absorbers

The MG TD came with either Girling or Armstrong shock absorbers at the front. Mine are Armstrong.

TIP

OVERHAULING ARMSTRONG SHOCK ABSORBERS

If Armstrong shock absorbers are being overhauled, it's important to make sure they are fitted with new, drilled bushes and seals for the lever arms. A cheap overhaul may involve just fitting new seals in the existing holes, which are often too worn to seal properly and will thus leak fairly quickly. It may increase the cost, but it's worth opting for new bushes in the long run.

▼ An Armstrong lever shock absorber that was sectioned to be used at motor shows by Armstrong themselves, in period. It shows quite clearly how the unit works. *(Moss Europe)*

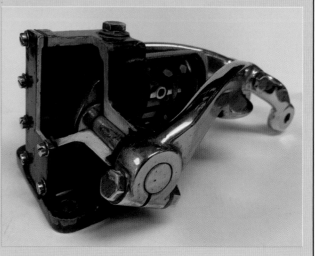

They look quite different to the modern single-piston units we see today, but the way they damp the spring energy is exactly the same. In my units there are two pistons that travel inside cylinders. The movement of the piston is damped by fluid, and it's this fluid that absorbs the energy. Modern dampers work on exactly the same principle, but the system has had a few decades of development to refine the way they work.

Dismantling

I had already removed the front suspension and steering components from the chassis in big chunks. Now they were off the car, I placed the assemblies on a big, open workbench and set about pulling everything apart as much as possible. I wasn't worried about snapping or stripping the odd nut and bolt, as I would be replacing all the nuts and bolts with new ones.

I decided not to dismantle the Armstrong shock absorbers, but took the top cover off to check the components inside, and have a closer look at how it all works – it's very cool. These units can be rebuilt, but they aren't easily available to buy, so I was careful with them. Also, one of the suspension upper arms is a press fit, so I didn't force it off, I left it where it was.

I dismantled the other upper arm and the lower arms, and also removed all the rubber bushes, as I would use new ones when the components were reassembled.

◄ **An advert for Armstrong lever shock absorbers.** *(Vintage and Classic Shock Absorbers)*

I only retained the following components:
1 Lower wishbone arms
2 Lower wishbones plates
3 Lower wishbone mounts
4 Uprights
5 Upper arms
6 Armstrong dampers
7 Bump stops (I used new ones when reassembling)

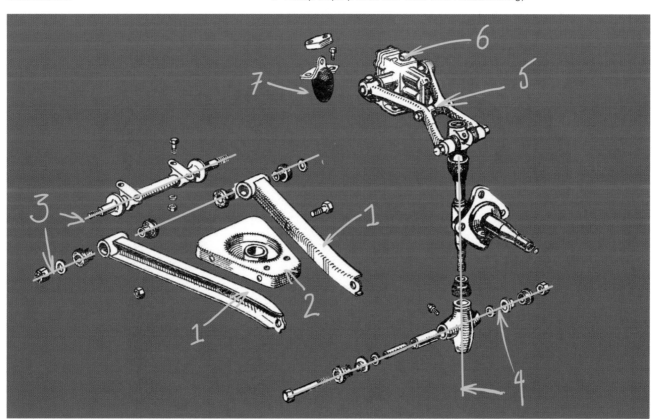

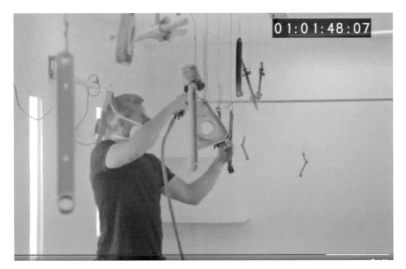

01:01:48:07

▲ **Be sure to wear the correct mask when in the paint booth. I hung the components with adequate space around them to paint.**
(Master Mechanic)

My parts were all really solid and reusable. If there is a corrosion or damage issue, it isn't worth taking any chances, as most parts are still easily available to buy – a local MG car club should be able to help with contacts.

Prime and paint

With the bare-metal parts cleaned and ready, I re-masked the areas that didn't require paint (the threads of the uprights and the stub-axle surface). I then used some decent acid etch primer in an aerosol 'rattle can', followed by a nice generous layer (or three) of gloss black paint. I am friendly with a local paint-booth owner who lets me drop in from time to time to tackle these jobs when his booth is free, so when the opportunity arrived, I quickly popped round to spray. With the booth heater on, the parts dried really quickly.

I didn't waste my time with the old drum brakes and hubs, and they joined the growing pile of unused donor parts that could be sold off later for tea and cake money!

Clean up

With the front suspension in pieces, I spent some time carefully degreasing each part. I have a small parts degreaser in my workshop, which made light work of it – a plastic bowl and some elbow grease works as well, but is just a bit slower. Once the parts were grease-free, I protected the critical areas (like the machined-finish stub-axle surface and threaded sections on the uprights) with strong tape, then I threw the lot in the blast cabinet to clean them back to bare metal. Again, I'm lucky enough to have a cabinet in my workshop, which is great for these smaller items, but this could, of course, be tackled by hand.

New parts required

The ethos of my build was to start as I meant to go on, and build a quality, new-finish vehicle, with care and attention to detail, so I replaced the following front suspension components:

1 New upper and lower trunnions – sometimes called kingpin bushes
2 New upper and lower rubber or polyurethane bushes
3 New nuts, bolts, washers and split pins
4 New fluid for shock absorbers
5 New rubber bump stops
6 New, or cleaned, grease nipples – this suspension system relies on being greased, so I needed to make sure the nipples were not blocked and actually allowed grease into the system.

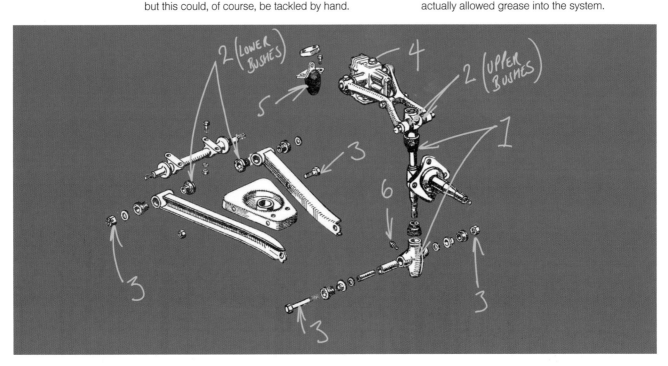

▶ **Rebuild kits are available for the MG TD front suspension. Mine had all the parts I needed, and a few I didn't!** *(Master Mechanic)*

Rebuild

I found a really handy rebuild kit that had all the parts I needed (and a few I didn't), and got all the parts laid out on the bench. Alternatively, the parts could be sourced separately.

As previously mentioned I won't go into suspension rebuild details, but will stay focused on the meat of building a special. If you need support for the rebuild details of donor parts, try the relevant Haynes Manual for your specific make and model, or have a peek online. The MG TD front end is pretty straightforward to bolt back together in any case.

Refitting the front suspension to the chassis

Protecting those threaded inserts at the blasting stage paid off here. As mentioned previously, I sourced new nuts, bolts (and spring washers) to use during reassembly.

First of all, I refitted the dampers to the chassis. All these newly painted and refurbished parts would eventually have to come off again (to paint the chassis), so it was OK to just hand-tighten the fixings for now, but it was important to properly tighten the shock-absorber fixings, although I fitted just two of the four bolts (the horizontal ones) to hold it firmly in place. It was essential to securely fix the dampers, as they would provide the datum point for the centreline of the front axle (the middle of the dampers have handy reference lines cast into the housings), and this would provide a reference point when measuring from the front-axle centreline across the rest of the car.

Next, I fitted the lower wishbones to the chassis using the four nuts and bolts, then fitted the uprights with the lower and upper trunnion bolts, bushes and end caps.

Finally, I fitted nice, new bump stops.

At this point I could stand back and admire the first bit of actual car building. Marvel at the double-wishbone action, as it keeps the stub axle perfectly level as the suspension moves up and down. I really admire the simplicity of this design, which is partly why I insisted on keeping the original front end from the MG.

For now, that was it for the front suspension. This may seem a little premature for a build that had just started, but it's important to me to 'plant' the four wheels when building a special. After all, everything else can be changed and will eventually fit in and around the wheels and suspension. So, with the front layout established, I turned my attention to the rear.

▲ **The Armstrong damper bolts back to the chassis really simply. The first bit of finished work, but still a long, long way to go yet.** *(Master Mechanic)*

◀ **The finished front-left suspension assembly. I am a massive fan of the double wishbone set-up.** *(Chris Hill)*

CHAPTER 7

Donor for running gear

At this stage of my build, I set about looking for a 'live' rear axle – one capable of taking more power than the modest MG. I naturally fell upon the Alfa Romeo Spider, which is relatively easy to get hold of in the US, fitted with rear disc brakes, and nearly every 2.0-litre Alfa Spider delivered to North America was fitted with a limited-slip differential (most of the smaller-engine cars were not). The bonus was that

it was made by Alfa, which felt like a rather nice nod to my build.

I found a rather sad-looking car in an advert online, while searching for 'Alfa Romeo Breakers', and like my chassis-donor MG TD, it's fair to say this car was no longer a viable restoration project. She was a little beaten up (that's an understatement), but as a donor for my special she had all the right ingredients. It didn't matter that she was missing a door – after all, I didn't need a door. Although I set out seeking an axle, I ended up getting a whole car that provided so much more, including the engine and gearbox, for my special!

The great thing about the Spider is that Alfa Romeo built and sold this car for three decades, and any car that lasts almost 30 years must be doing something right. In the interesting and ever-changing market for classic cars, unlike many Alfas before it, the Spider has yet to hit crazy prices, making it a sensible and affordable donor.

It was sensible for me to use one as a donor, as the USA was a massive market for the Alfa Spider, despite the car enduring various regulatory-based atrocities, such as girder-like bumpers and a raised ride height (another MGB parallel there). Because of that, there are still plenty of poor examples available for sensible money. I'm a great fan of the model, and I loved the Spider I rebuilt on *Wheeler Dealers*. This

▼ **The post-1967 Alfa Romeo Spider shot for the car's launch publicity in the style of the time.** *(Centro Documentazione Alfa Romeo – Arese)*

▶ **The example I found was in slightly worse condition, but it did have all the parts I needed.** *(Ant)*

CHECKING THE DONOR ENGINE

With a car that has sat for a while, the first move is to try and turn the engine by hand before cranking it over. Using a socket on the crank pulley bolt, rotate the crankshaft, checking everything moves freely and isn't seized. If it feels really tight, pull a couple of the spark plugs out and try again – it might just be gummed up due to sitting for years, and struggling to turn against the cylinder compression.

As it's likely that the engine will be stripped and rebuilt anyway, this check might be all that's required. If the engine turns, it's probably a rebuild candidate, but if the engine is to be started to see if it will run, it's best to take the following steps:

- Check the state of the distributor cap and rotor arm – the condition of these might give an indication of the engine state when it was last parked.
- Check that the plug leads are not obviously broken or damaged. They don't have to be brand new, and as long as they transmit enough voltage to produce a spark, they will do for now.
- Check there's oil in the engine. Try to get a feel for the state of the oil – does it have grit in it? Is there any white goo under the filler cap where oil has mixed with the coolant?
- Does the engine have any coolant in it? If it doesn't, you should be fine starting the engine

◀ If the donor engine is going to be completely rebuilt, you really only need to check that it turns freely and isn't seized. *(Chris Hill)*

for a second or two, but no longer. Certainly don't take the car for a drive.
- If there is a battery handy, rig it up with jump leads and just turn the starter for a second. Don't crank the motor for long, just check it works and the wiring doesn't catch fire (that has happened to me before!). Get her turning over bit-by-bit on the starter motor, and it will soon be obvious if anything has seized.

car, however, needs to be handled with a different mindset – this is a donor, not a restoration project.

When selecting a car as a donor, it's imperative that each and every component that is likely to be used is carefully checked. In my case I was lucky enough to get in and drive the car, proving the engine not only turned over, but also ran. The axle held oil, drove without noise, and the brakes worked really well. Perfect!

In some cases, the wreck may have been sitting for some time and will need a more gentle approach. It always pays to get as much information from the seller as possible. How long has she sat? Does she drive? When was the last time? Any known issues, and so on...

If the potential donor starts and runs OK, it might be worthwhile taking it for a quick drive if it's safe to do so. Any knocks, rattles and other odd noises should soon be obvious, but before setting off, always check the brakes!

My Spider was missing a propshaft support bearing and had no clutch, but I really wanted

to check the rear end was solid and not making horrible noises. I drove it around the block by snatching second gear when I could. Brilliant fun! It was at this point I knew the whole car was coming home with me.

With the purchase secured, and the Spider safely transported back to the workshop, I was champing at the bit to remove the parts I needed. At this stage my aim was simple – to remove the three big chunks I knew I needed from the donor: the engine, the gearbox and the rear axle.

Removing the engine and gearbox

Pulling engines out of cars can seem daunting, but it doesn't need to be. It just needs logical steps, taking care of one system at a time – electrical, hydraulic, cooling, etc – and moving on from there.

I've removed countless engines from all kinds of cars, and in more complicated cases it can be helpful to label things along the way, but this '70s Alfa is about as simple as it gets. It was

▲ **Out she comes, slowly but surely. Removing an engine like this involves less work than you might think.** *(Master Mechanic)*

▼ **The trick to making sure things go smoothly is being methodical and taking your time.** *(Master Mechanic)*

▼ **The engine may look tatty here, but it won't once I'm finished.** *(Master Mechanic)*

pretty obvious what went where, and anyway, the engine was never going back into the donor car, so it didn't matter if I forgot where the windscreen-wiper electrical connection went. So, on with the dismantling.

I started by removing the battery, then I disconnected the wiring at the engine. There was no need to trace all the wires back and disconnect at the other end – I just disconnected whatever was connected to the engine.

Next, I removed the gear stick. This was the only component to remove from inside the car, so I first removed the gear knob and centre console, then the lever itself.

At this point, I drained the coolant. This must be disposed of properly – animals think it's really tasty, until it poisons them. It must never be tipped down the drain. The rubber coolant hoses were obviously going to be replaced, so I didn't need to be too gentle with them. I just disconnected them any way I could, but it's important to protect the metal surfaces on the engine as this is done.

Next, I removed the radiator.

The fuel line was next, which is usually a simple connection. The tank must always be drained first, into a proper fuel container.

In essence, that was all of the life support that fed the engine disconnected or removed. Next up were all the hard connections.

Exhausts can often be a pain to remove. They sometimes stay in place for years and can get rusted on due to the rain – especially in areas where the roads are salted – so it may be necessary to be brutal here. It might even take a reciprocating saw to remove the exhaust at the manifold.

Next, I unbolted the propshaft at both ends and set it to one side, as I would need to reference it when the time came to get a new one made to a bespoke length.

There was now nothing attached to the engine at all. It was just a lump sitting in the car, and now was the time to start being a bit careful. The engine is heavy – the Spider's weighs over 90kg (200lb).

It was now time to remove the gearbox mounting, and then the engine mounting bolts/nuts. The engine/gearbox was now sitting in the car, but not bolted in, so an engine hoist was essential to lift the assembly out at this point.

I attached a bracket to the top of the engine cylinder head, and slowly took up the tension in the chain. The engine lifted away easily, but occasionally the whole car may start to lift instead, in which case something is still connected, or caught up, so it will be necessary to lower it back down and recheck everything.

Once the engine started moving, I wrestled it up to the top of the engine bay. I had to angle the engine to allow the gearbox to come out as well. There was no need to remove the gearbox separately, as there was sufficient space to pull it out joined to the engine as one big lump.

That was it. Once it was clear I pulled the hoist back and lowered the engine and gearbox gently to the floor. Dropping it onto a nice-sized wooden pallet is often a good idea.

Removing the axle

Removing the axle was similar to the engine. I didn't have to be too careful about the parts I wasn't going to keep, which made things easier. If this is being done on the ground, rather than on a lift, it will need to be approached with a certain degree of caution, as the whole weight of the rear of the car, springs under load, and shifting forces will come into play.

If this job is being done without a lift, it will be necessary to jack the car up and support it on axle stands, placed at suitable points under the body of the car. The wheels and brake calipers will then have to be removed, at which point the rear axle will be hanging down, with the limit straps holding the weight. Coil-spring compressors will be required to remove the springs, and this must be done very carefully, as they hold a lot of spring energy. The lower shock absorber bolt can be removed, or the dampers can be removed altogether.

At this point, a trolley jack can be used to just take the weight of the rear axle so the limit straps go slightly loose. The limit straps can then be cut – I used a reciprocating saw. The bolts that pass through the rear leaf spring can then be removed, at which point the axle will be balanced on the trolley jack. The assembly can be carefully pulled out from under the car using the jack.

My car was now immobile. I simply lowered the car onto one of the rear wheels and tyres that I sat on top of a 'go-jack', and then pushed the car out of the workshop. It was funny how light it was now. The remains lived outside, as I knew there were more parts to salvage from her. I wasn't going to send her to the scrap heap just yet – this donor would come in handy later.

I was excited at this point, so I placed the axle (as it was) under the rear of the MG chassis (using a jack to help manoeuvre it) and then, using an engine crane, hovered the engine and gearbox in place too. It's important to pause for these moments...

Notice that I marked my intended layout and the wheel positions on the floor using masking tape (well I am a big kid!). It gave a sense of the size and proportions I was working with, and showed that single-seater does not mean smaller.

For now though, I remained focused on the next step – the rear axle.

▲ **Instead of unbolting them, I simply took a saw to the check straps – and I'm not even sorry!** *(Master Mechanic)*

▼ **Taking a first look at how the engine might sit in the chassis. Everything is fluid at this stage, and nothing is set in stone.** *(Chris Hill)*

CHAPTER 8

Rear axle

▼ **Fangio also sat well clear of the top of the car, as can be seen in this picture taken at a wet practice session at Silverstone, in 1950, which also shows he wore the clothes to match his car's Argentinian blue and yellow-coloured nose band identification (or moustache as custom-car designer Chip Foose calls it). A small detail I love for my car.** *(Guy Griffiths Collection)*

My Alfa Romeo Spider rear axle provides the first chance to get really creative with some ambitious re-engineering for my build. Numerous road cars utilise a 'live' rear axle, so the choices out there are endless. Obviously, having a single donor car for the running gear assists me in my build.

As it stands, the Spider axle has a track width of 50in (1,270mm), and the differential is located somewhere just off-centre. I not only want to increase the track width to 52in (1,321mm), to match the wider Alfa 158 rear track, I also want to completely relocate the differential.

Five-time World Champion Juan Manuel Fangio had huge success in the 158. He was a racing driver, and like jockeys they tend to be rather small. He was quite short at 5ft 9in, compared to me at over 6ft 2in, and even Fangio looked rather silly poking his head out the top of the 158 in period – there, I said it! This is special building – my car, my way. I wanted to sit low and comfortable within the finished car. Not only would this create a better seating position, but it would also lower the centre of gravity of the car and driver. After the engine and gearbox, and I hate to say this, I am the second-heaviest part of the whole

LIVE AXLES

A 'live' or beam axle is a solid or rigid casing containing a differential that is usually mounted somewhere near the centre of the assembly. The axle transmits drive to the wheels via two half-shafts, and was patented by Louis Renault in 1899. This arrangement was common on rear-wheel-drive cars up until the 1980s, and is still common on small commercials. In general terms, the ride and handling characteristics it provides are not as sophisticated as those enjoyed by cars with fully independent suspension because, in simple terms, the axle is a rigid unit – if one side of the axle goes up, the other side tends to go down. However, when well located, a live axle can provide good and, most importantly, predictable handling. Also, although it may lack the last fraction of grip compared to an Independent Rear Suspension (IRS) back-end, a live-axle car is often actually more fun to drive than an IRS car, and this special is about fun!

Additionally, live axles are easy to engineer, which is a bonus for me. In its original form, the real 158 used a swing axle and transverse leaf-spring rear suspension, with both friction and hydraulic dampers. It later adopted a DeDion tube arrangement. Both require heroic levels of driver skill and although I may *want* to be Fangio, I know that I'm not...

vehicle. The lower I placed all that weight, including me, the better the car would handle.

Because the propshaft would normally pass right up the centre of the car, meaning that the driver would have to sit above (astride!) it, if the differential is offset (which will also mean that the propshaft is offset), it means that the driver can sit lower in the car, so lowering the centre of gravity, and that's exactly what I wanted. Aside from better weight distribution, offsetting the diff also helped with a number of other engineering issues. Moving the propshaft and gearbox to one side placed the existing Spider gear stick within reach of my right hand, rather than between my legs. It also meant that the three pedals could be grouped together, whereas with a centrally mounted differential I would have to split them, as I would be forced into straddling the gearbox, meaning I would need to place the clutch pedal on one side and brake and accelerator pedals on the other

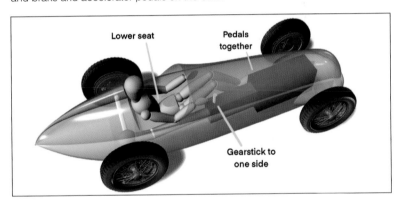

Lower seat   **Pedals together**

Gearstick to one side

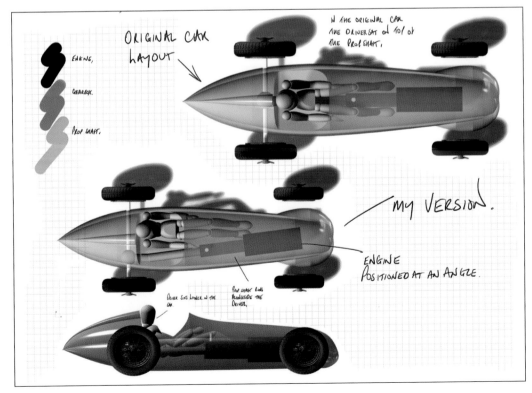

▲ **Getting down low. In order to sit lower in the car than Fangio did in his Alfa 158, I had to make some pretty big decisions.**
(Paul Cameron)

◀ **Paul's drawings show how my drivetrain and control layout differs from the original 158. You can see how, with careful planning, it's possible to adapt details and make compromises to personalise your car to suit your needs.**
(Paul Cameron)

(common in period race cars, but less than ideal for me and what I want from my build).

This arrangement may seem ambitious, because it is! I had to take my time and work everything out carefully.

The expensive part of the Spider rear axle is the differential unit, and having a bespoke one built would cost a fair amount. Using a matched driveline, with the diff, gearbox and engine coming from a single donor, is the best solution for a car builder.

I retained only the aluminium diff casing (and internals) and the inner portion (the flanges) of the steel axle tubes that bolt directly to the casing. The rest was built from new. First, I dealt with the axle casing itself, before moving on to the rear suspension set-up.

I started by stripping the axle down to component form. The internals could be safely stored for now, as we were working solely on the casing.

Using an angle grinder, I cut both the inner flange sections off the Spider axle – the parts that bolted to the aluminium diff casing. The remaining part of the axle casing was discarded. There was a weld line where Alfa Romeo welded these flanges to the axle tube in the 1970s, and that weld line was exactly where I cut.

▲ You can't make an omelette without breaking some eggs! I had to get pretty brutal with my rear axle. *(Master Mechanic)*

▶ The bits of the original axle that really mattered were the central diff casing and the flanges. The rest got replaced. *(Ant)*

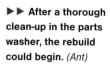

▶▶ After a thorough clean-up in the parts washer, the rebuild could begin. *(Ant)*

Originally, I was going to retain the existing Spider axle casing, cut off all the Spider brackets and perform the offset by altering the original tubes and using the existing Alfa outer bearing housings. Instead I opted to replace all these parts for new.

I swapped the bearing housings for what are known as 'Ford small bearing' units. These bearings are slightly larger and more robust than the Alfa's, and easy to get hold of, plus the choice of rear brakes to match the Ford units is vast, affordable and far easier to get hold of than the Alfa parts. So, I used Alfa inner and Ford outer units.

Making the new axle casing

The new axle casing tube was made from 2½in mild steel with a 6mm wall – heavy, thick-walled and strong – a great starting point for a new race-car axle. First, I worked out what I needed in terms of components, and then it was time to ready those components.

I cleaned and shot-blasted my recycled inner flanges and laid them out. I then cut two pieces of the tube – one 660mm long and the other 275mm. Next, I ground down the outer edges of each tube to create a 'valley' to weld into later, and again laid them out. These seven components are all that is required to make the complete new axle casing.

◄ **Ford small bearing housings are perfect for the rear-axle outer bearings.** *(Ant)*

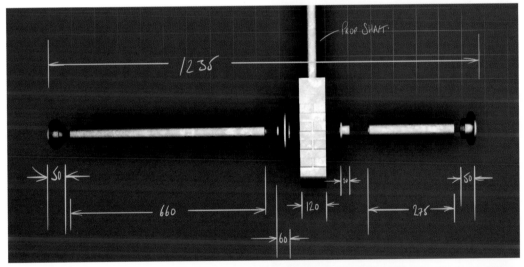

◄ **Measure twice, cut once. I must have checked and rechecked these dimensions 100 times.** *(Paul Cameron)*

▼ **The Magnificent Seven. These parts are all that are required to make a brand new 'offset' axle.** *(Ant)*

▲ **TIG welding is the most friendly form of welding for this. It's very precise, doesn't leave any mess and the heat is very controllable.** *(Master Mechanic)*

▼ **The axle casing is assembled around a perfectly straight rod, made specifically for this purpose.** *(Master Mechanic)*

▶ **Three passes of the TIG welder on each joint. Engineering projects from submarines to gas pipes have multiple weld passes – any environment where failure is not an option.** *(Ant)*

The axle casing is merely a shroud to house the inner workings of the axle (the differential and driveshafts). It is essential that the casing components are welded together accurately. The aluminium diff casing and the bearing housings at each end have to be perfectly aligned, as the driveshafts fit between them. If the components are not precisely aligned, the rotation of the wheels will destroy the axle (or likely the bearings) within minutes.

To ensure accurate alignment, a jig must be used. I used a precision-ground alignment rod, sold specifically for this purpose – it's a long, incredibly stiff, thick-walled tube, which is machined to be perfectly straight. This forms a centre spine for the tubes to follow, which keeps everything correctly aligned.

The rod works in conjunction with some extremely precise pucks. These slip over the alignment rod and keep the individual components centred. A very close second-best is building up layers of masking tape around the rod until the components fit snugly. Layers of tape can be wrapped around the rod, until the correct internal diameter is reached, then the components can be slipped over the tape, keeping them central to the rod, and held firmly in place using grips.

I assembled one side at a time, passing the three components over the rod using the pucks to keep things aligned. I then tack-welded the components in place, making sure that I rotated the outer bearing housing so that the flat edge sat at 12 o'clock when the assembly was bolted to the diff casing. The positioning of this is important, as it determines the finished brake caliper location.

It was essential to tack weld several times in various areas without putting too much heat into the components – the tacks would hold it all square until it was time to fully weld. When welding, I opted for three passes of a TIG welder across each joint, allowing time between passes for the metal to cool.

I repeated the procedure for each side, and the key to getting it right was preparation. I had to ensure that each loose component sat firmly on the rod and had no chance of slipping or moving – it had to be square.

Once both new tubes had cooled down, I temporarily refitted them to the differential casing, using a handful of nuts, and firmly tightened them. Note that the flat tops of the outer bearing housings are located at 12 o'clock on each side, ensuring accurate brake-caliper location.

I then used axle stands to position the axle underneath the MG TD chassis, setting the axle so it sat in line with the new 158 rear-axle centerline (marked in pen earlier in Chapter 5, 98.6in from the front axle). It was important to have half an eye on the ride height at this stage, so I made sure the

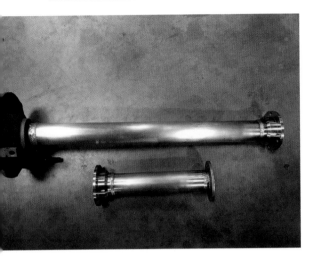

▲ The finished items. If ever there was a time to go slowly and be sure everything is precise, this is it. *(Ant)*

▲ Bolted back together, the finished unit looks like it could have been made in Turin. *(Ant)*

▼ The fruit of my labour. The finished axle assembly offered up to the chassis and supported on axle stands. *(Ant)*

chassis was 13¼in from the floor, and the centre of the axle was 14in off the floor to simulate the centre of the rear wheels. At this point I noted the new position of the differential unit, and visualised my posterior sitting snugly to the left of it! Pretty cool, huh?

Locating the axle on the chassis

Now the casing was pretty bare and required some additional locating components to fix it to the chassis. I did this with the axle resting in position in the chassis, and I opted for the tried-and-tested combination of four trailing arms and a single Panhard rod.

Fixing the trailing arms to the axle required some bracketry. I saved myself some time by taking my own hand-drawn sketch to a local company with a CAD water-jet cutter. Made from 4mm mild steel, the brackets ('boomerangs') are pretty hard to cut by hand. Local laser cutters (or in my case water-jet cutter) are now common, and finding one nearby should be easy and will save a bag of time, as the car requires four of them (I got a fifth one cut to act as a jig later). It should be possible to cut these by hand, but it would require a lot of determination, and I'd love to see photos!

I assembled the rear suspension using ½in rose joints all round – five right-hand threaded and five left-hand threaded, with locking nuts for all ten. I also used some mild-steel, threaded 'mushroom' inserts that matched the same ½in thread as the rose joints. I used 1¼in-outer-diameter inserts for the trailing arms and Panhard rod, and 1in-diameter for what would become the damper mountings. I could have made these from scratch – however they were really easy to source online. Again, five left-hand-threaded and five right-hand-threaded inserts were needed.

▶ **My hand drawing of the 'boomerang', converted to CAD.** *(Paul Cameron)*

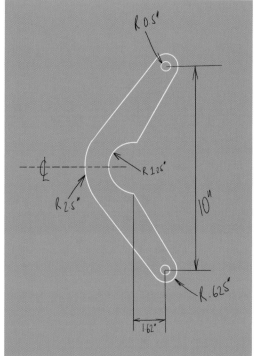

▼ **Time to get friendly with your local water-jet (or laser) cutter. You could cut these by hand, but it's just not worth it.** *(Master Mechanic)*

▶ **Rose joints are not expensive, and perfect for the trailing arms. I used ½in joints.** *(Ant)*

Next, I bolted two of the 'boomerangs' together, with a rose joint in place on each bolt between the boomerangs, then securely tightened the nuts and bolts to fix them together. I needed two of these assemblies, one for each side.

Now it was time to cut four 17½in-long pieces of 1¼in-diameter steel tube to make the trailing arms. I opted for 14SWG (Standard Wire Gauge) CDS (Cold-Drawn Steel) tube. There are a number of ways tube can be manufactured, but seamless is the strongest and most dimensionally accurate. The build requires the mushroom inserts to be welded into the tube, and if there was a seam inside, it could cause problems. Plus, my car's suspension would be working under heavy load (my driving style), so I wanted the best strength-to-weight ratio possible, and that's found with seamless tube.

I inserted a left-hand-threaded insert into one end of the first tube, and a right-hand-threaded one into the other end, taking care not to mix them up. Each trailing arm required both thread types, which would later allow the ability to adjust each arm's length by simply twisting the tube. I tack-welded the mushroom inserts in place for now, added the

▲ **Mushroom inserts are easy to find and buy online, but you could make your own using a lathe if you really wanted to.** *(Paul Cameron)*

▼ **Make sure you don't mix up the thread directions on the inserts. One left-hand and one-right hand thread makes an easily adjustable trailing arm.** *(Ant)*

▲ The finished, assembled trailing arm unit. Very pleasing. *(Ant)*

◀ The inserts slide into the tube and are TIG welded in place to hold the adjustable rose joint. *(Ant)*

locking nuts to the rose joints, and screwed the rose joints into the arms so they were threaded in to a depth of half their thread length (this allows maximum adjustment each way without running out of thread).

I then placed each boomerang around the relevant new axle tube to make sure that it all sat square to the floor and lined up (as per the accompanying illustration and photos).

Once I was happy with everything, working on each side of the axle assembly in turn, I tack-welded

◀ Forgive my mix of imperial and metric measurements... Keeps you on your toes! *(Ant)*

▼ I'm indicating a right-angle here for the TV camera, while resting my busted arm!
(Master Mechanic)

▼ The boomerang slotted around the axle tube and was welded on perfectly square. *(Paul Cameron)*

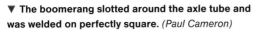

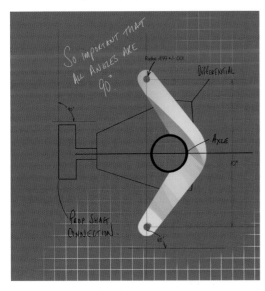

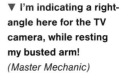

▲ Two boomerangs welded side by side, with the trailing arm in the middle. *(Master Mechanic)*

▲ Welding with one arm is possible, but it's easier with two. *(Master Mechanic)*

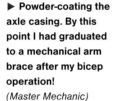

▶ Powder-coating the axle casing. By this point I had graduated to a mechanical arm brace after my bicep operation!
(Master Mechanic)

▼ Originally, I actually opted for the longer Panhard rod shown here, and I changed it when the body arrived.
(Master Mechanic)

the boomerangs in place on the axle tube using four hefty tacks. I then cut a piece of 4mm x 30mm (to match the width of my rose joints – ie, the gap between the boomerang halves) mild-steel flat bar to bridge the gap between the boomerang halves at the rear, in order to strengthen them. I placed a curve in the flat bar to match the rear profile of the boomerang and scalloped the ends so they didn't foul the moving rose joints. This makes these bespoke brackets incredibly tough, and they need to be, as they are taking all the cornering, acceleration

and braking forces at the rear end of the car. These are some hard-working components.

The rear boomerang set-up also provides the lower mountings for the spring/damper units, known as coil-overs (because the coil springs fit over the dampers). To provide these lower mounting points, I used 1in-diameter mild-steel mushroom inserts, with a ½in right-hand thread, welded to the curved strengthening plates at the rear of the boomerangs.

To do this, working on each side of the axle in turn, first I drilled a 20mm hole in the centre of the boomerang strengthening plate. I then screwed a bolt into the insert and left it in place to protect those threads, and also to assist in eyeballing the insert and ensuring it was welded squarely in place.

I placed the insert concentric with the previously drilled 20mm hole in the boomerang strengthening plate, and tack-welded it in place. It was important to ensure that the damper-mounting bolt sat at 90° to the axle and floor. I repeated the whole procedure to fit the spring/damper lower mounting on the remaining side of the axle.

Next, I used another ½in threaded mushroom insert to make the mounting point on the axle for the Panhard rod. I tacked it to the front of the axle tube, in the centre of the left-hand boomerang location (directly opposite the spring/damper mounting) and welded in place two triangular support gussets.

That was the axle casing done! I then fully welded it up, which I tackled one-handed... well, gotta keep moving!

I made sure that the tack welds were substantial enough that I could progress with the next phase of the build with the axle components just tacked together.

Once fully welded, I stripped the axle down and took my two steel casings to a local powder coater. They were shot-blasted to key the surface, and then I chose a classic gloss-black finish. I left the diff casing in bare aluminium.

Driveshafts

Now for the part of the rear axle I haven't mentioned yet – the actual driveshafts themselves.

In the past (my maverick days of one brave man in a cow shed), I modified axle shafts myself using machined sleeves to 'cut 'n shut' them together. This meant slipping a very close-fitting steel sleeve over the cut axle shafts and welding all the seams together, then drilling and pinning them for belt-and-braces safety. For this build I invested in some outside help. The axle shafts are subjected to a huge amount of rotational shear force, and while a cut 'n shut method will work, it is not ideal by a long stretch, and when you consider the speeds at which these shafts rotate, they also need to be precisely balanced.

I took the drawing shown (along with the axle, old driveshafts and differential unit) to my local axle specialist and got a pair of shafts properly machined, hardened and balanced. I decided it was worth investing in this professional service, and I didn't regret it. Any imbalance, however minute, will be felt in the finished car.

I simply slid the finished axle shafts into the rebuilt diff, but I didn't worry about tightening the fixings at this stage, as that would be done when the rear brakes were fitted.

With the rear axle itself completed, it needed to be fixed in place to work in harmony with the chassis, so the next job is to tackle the rear suspension.

▲ It's worth spending the time to properly overhaul the differential unit and fit new seals and gaskets. Once completed, the powder-coated axle looks fantastic. Note here the trailing arms and Panhard rod are connected to the chassis – on to that shortly. *(Chris Hill)*

TIP

At this point, it's a good idea to take some masking tape and leave a note on the diff as a reminder to fill it with oil later – a diff run with no oil will wreck itself remarkably quickly!

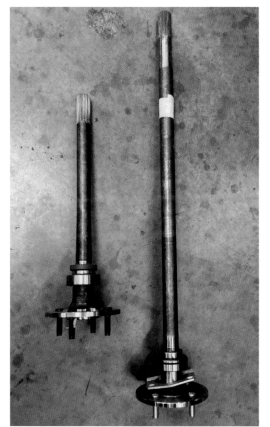

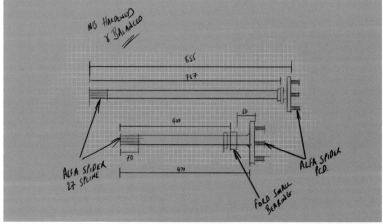

▲ There are not many axle shaft building companies out there. When you find one give them as much information as you can. *(Paul Cameron)*

◄ The finished rear axle shafts, ready for fitting. *(Chris Hill)*

CHAPTER 9

Rear suspension

With the trailing arm positions set on the rear axle (via the boomerangs) I now needed to make the remaining mounts to secure the suspension to the chassis. For this, I machined four bushes, each 1in diameter, with a ½in hole drilled through the middle, and 3½in long. To make the bushes, I used a piece of 1in solid mild-steel bar, cut to length, and then drilled out the centre on a lathe with a ½in drill bit. I also sourced four 6in-long bolts with nuts.

Next, I positioned the axle accurately under the rear end of the MG TD chassis, using axle stands. I positioned the axle so that it was:

■ At the correct ride height (14in from the floor to the vertical centre of the hub simulating the centre of the hypothetical wheel)
■ Positioned to give the correct wheelbase (I marked the rear-axle-centre position earlier, 98.6in from the centre of the front axle)
■ Rotated to ensure that the damper lower-mounting bolts on the axle sat at 90° to the axle and floor

With the four trailing arms fitted to the boomerangs on the axle, I bolted the newly made 3½in-long bushes onto the rose joints at the forward ends of the trailing arms, using the four 6in-long bolts and nuts, with the bushes positioned on the inner sides of the trailing arms. As there was a left-hand-thread rose joint at one end of the trailing arm, and a right-hand-threaded one at the other, with the locknuts loose, the trailing arms could simply be rotated to

▼ I made my chassis bushes on a lathe. You may be able to find some very thick-walled tube to do the same job. *(Master Mechanic)*

▼ The finished rear axle assembly, with boomerangs and trailing arms fitted, resting in place under the chassis. *(Master Mechanic)*

adjust the length between the rose-joint eyes. The distance should be 22½in, so I adjusted the trailing arms to give this measurement, and then tightened the locknuts to securely lock it all in place.

Now back to that additional boomerang that I had cut to act as a jig. Now was the time to use it, so I bolted the boomerang to the front of one of the trailing arms, using one of the 3½in-long bushes, plus a bolt and nut. With the axle correctly positioned, I could then clamp the boomerang to the chassis to provide the correct position for the bush on the chassis, and operate hands-free. Unfortunately, was tackling this portion of the build with an injured arm.

Working on one side of the chassis, and starting with the lower trailing arm, I placed a spirit level on top of the trailing arm, so that it was in a horizontal position, then I clamped the boomerang to the chassis, to give the correct position for the bush. It hovered 200mm below the chassis (see photos on the right). For reference, the photo shows the dimensions for the correct position of the bush. If things don't look quite right, it always pays to recheck the measurements.

I now had to make brackets to locate the trailing-arm front mountings. The MG TD chassis rails are 3½in wide at this point, so using some 3½in-wide, 4mm thick flat bar, I cut two pieces 6in long and welded them between the chassis and the bush to triangulate it (I used the spare boomerang clamped to the chassis to hold the lower bush and trailing arm in place).

I made sure that the bush sat flush with the inside edge of the chassis (so it didn't intrude into the cockpit) and the excess protruded outside the chassis. It was also important to ensure the bush sat at 90° to the trailing arms, and the floor. This may look odd, as the MG TD chassis tapers, widening at the rear, and this means that the bush won't sit at 90° to the outer wall of the chassis – it will, however, be at 90° to the axle, and that's what matters.

Once I'd tacked the bushes in place, I made some triangular cardboard templates to fill the sides, then transferred to some 3mm mild steel, and made plates to fill in both sides to provide strength. I fully welded the plates, then tickled them with a grinder so they looked as if they were part of the original chassis. I repeated the procedure on the other side of the chassis, and that was the lower trailing-arm mounts done.

The upper trailing-arm mount location was a twist of good fortune. Had I wanted the upper trailing arms, like the lower ones, in the parallel position, I would have needed to use more steel plate to make a second bracket to raise the bush above the chassis. However, I decided to angle the upper trailing arms downwards, because I wanted to add something called 'instant centre' to my suspension

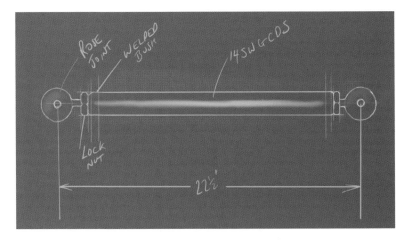

▲ **The Panhard rod is made in the same way as the trailing arms. It prevents the axle from moving sideways, while still allowing suspension travel.** *(Paul Cameron)*

▼ **Ideally, you'll have a very flat floor and can level your chassis accurately.** *(Chris Hill)*

▼ **I'm welding on the bushes with no gloves here – not best practice! Be sure to follow appropriate PPE recommendations for each job.** *(Master Mechanic)*

► **I turned the chassis upside-down for access, so I could get a full MIG weld around each edge of the bracket.**
(Chris Hill)

▼ **Again, the camera caught me not wearing gloves. Please do as I say, not as I do – wear the PPE!**
(Master Mechanic)

And now for the 'instant centre'. Enter orange string...

Instant centre

Instant centre is a theoretical point in space, where both trailing arms would meet if they were extended forwards. All cars with upper and lower trailing arms at the back should have the trailing arms positioned so that they would eventually converge. If the two trailing arms are perfectly parallel, that is known as 'infinite instant centre' (ie, draw a hypothetical line along each arm and those lines would carry on forever without meeting).

For my car, adjusting the length of the upper trailing arm dictates that meeting point between the two lines, and that instant-centre point will have a big impact on how the car handles under acceleration, specifically when it comes to maintaining traction.

Drag racing teams pay a lot of attention to instant centre, and they know a thing or two about transmitting power to the ground! A drag car is a good case study to explain the theory. The engine power comes down the propshaft, through the differential and out through the driveshafts to the wheels. As the wheels rotate, the car actually wants to lift slightly, and you see that, as drag cars sometimes tend to pop a wheelie off the start line. That lifting action is caused by forces travelling through the suspension components – in our case the trailing arms – and those forces can be used to our advantage (see diagram at the top of page 67):

■ **1** When power is applied, the differential wants to rotate in the opposite direction to the wheels
■ **2** The top trailing arm is under tension – it is being pulled by the differential as it tries to rotate
■ **3** If the trailing arm is angled downwards towards the chassis, then it will try to pull the chassis upwards slightly as the differential tries to rotate in the opposite direction to the wheels
■ **4** In turn, this forces the back end (the wheels) into the ground, providing more grip

geometry. This meant I could simply weld the upper bush to the top of the MG TD chassis rail.

I lowered the upper trailing arm so it rested on top of the chassis rail then, ensuring the upper bush again sat at 90° to the arm (and not the chassis rail), I welded it in place.

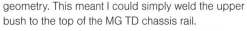

► **The instant centre is the theoretical point in space where the two trailing arms would meet, if they were extended forwards.**
(Paul Cameron)

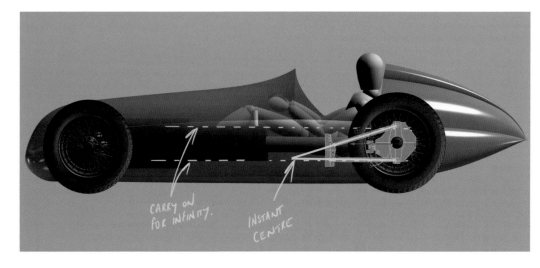

▶ **The diff wants to rotate in the opposite direction to the wheels. That force can be used to improve grip under acceleration.** *(Paul Cameron)*

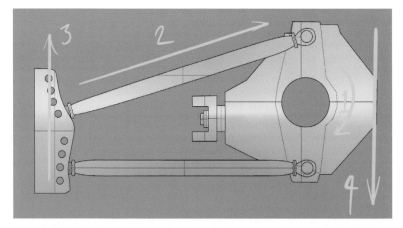

Imagine standing next to a table and pulling upwards on it. Your feet will be more firmly planted, as the weight of the table gets added to your own. If you stood on a weighing scale, you would see the weight increase. That weight increase improves friction between the tyres (your shoes) and the ground. More grip, less slip, better acceleration – win, win! And all this happens for free, just with a bit of clever geometry.

A regular road car will have an instant-centre point somewhere in front of the front-wheel centreline. A top-fuel dragster capable of getting to 100mph in under a second might have an instant centre positioned only a few inches in front of the differential, really harnessing those forces.

That level of aggressive traction force just isn't required for a road car, or my 1930s-styled track car, but if covering 400m in less than 4 seconds is your game, it's very important, and proves that instant centre really is a point to consider.

Using string lines attached to the trailing-arm mounting bolts on the boomerang, I brought the two pieces of string together to position my instant centre just behind the front-wheel centreline, and when I traced it back, I found, very luckily, that the front end of the upper trailing arm landed with the metal bush resting on top of the chassis rail. Result!

That's as far as I took the rear suspension for now. For my build in real time, I moved on to some further chassis modifications next...

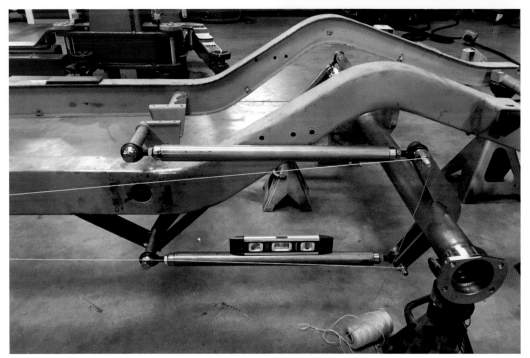

▲ **Drag-car builders use the principle of instant centre to improve grip, as Brittany Force shows..** *(NHRA)*

◀ **In this photo I've positioned the lower trailing arm horizontally using a spirit level, and because I've clamped my spare boomerang to the chassis to hold the front ends of the trailing arms, both arms are parallel. If I'd kept the upper trailing arm parallel, then of course it would hover above the chassis rail as shown here. Here, I'm using the string to position the instant centre.** *(Master Mechanic)*

CHAPTER 10

More chassis work

I already knew that the rear of the chassis needed to be removed, as it would stick out of the boat-tailed body shape. Up to this point I had left the middle crossmember in place to help retain some chassis strength in readiness to make the desired rear-end modifications.

Now it's at this point that I have to hold my hands up! If you watched the TV series covering the build of this special, you will have seen I made a bit of an error. Whoops! I built my body in the UK, while building the chassis in the US. When the body arrived it didn't fit...

This meant that I had to redo my work on the rear end! In this book, I will of course skip my mistake, call it research and development, and progress with the correct path, while whistling as if nothing happened! Here we go...

▼ **Nothing to see here. The body fitted first time, honest guv'nor.** *(Master Mechanic)*

◀ **The section from the back of the MG TD chassis that I didn't need. Nice to finally get rid of that.** *(Chris Hill)*

▼ **Cutting into the rear chassis just behind the suspension mounts to allow me to taper the rear end.**
(Master Mechanic)

I measured 94.5in back from the front axle line, and marked on the chassis with a pen. Using an angle grinder, I cut the rear end of the chassis clean off.

Next, I made some thin slices into the chassis, just behind the newly fitted upper suspension mounts. I only cut the inner, upper and lower chassis wall, taking care not to cut into the outer sidewall of the chassis rails. I tapered the cut from 4mm to 0mm with a fine cutting disc, and that worked perfectly.

The rear of the chassis was now fairly flexible, allowing me to move the chassis rails inwards, tapering the rear end so it was narrower. I cut a new rear crossmember, 25¼in long from 3in x 2in mild steel box-section tube – 16SWG ERW (Electric Resistance Welded) tube is perfect. I tack-welded it in place so it sat flush with the top of the MG TD rails.

Using a tape measure, I measured the chassis diagonally, corner-to-corner, to ensure the rear end of the chassis was square. The slices made in the chassis rails earlier were now closed up, and I welded them fully to secure the rear end (again, making sure it was square).

Having tacked it earlier, I then fully welded the 3in x 2in crossmember to the chassis rails. The

▶ **The tapered cut, ready to move the rear chassis rail inwards. Note that I used fireproof welding sheet to protect the fresh powder coat.** *(Master Mechanic)*

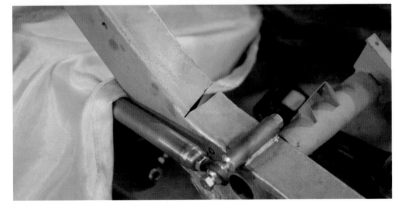

▲ Checking the fit of the new rear crossmember. Suddenly progress was starting to accelerate. *(Master Mechanic)*

▶ The finished modified rear end of the chassis, with the chassis rails tapering inwards, and the new rear crossmember welded in place. *(Chris Hill)*

chassis now tapered to the trailing arm mounts, before tapering inwards even further to fit within the tail.

I added some 1.5mm steel end plates (which could have been made from whatever flat steel happened to be kicking around) to close off the sides of the rear crossmember, and then fully welded them in place.

At this point I removed the middle crossmember (left in place to strengthen the chassis until it was finished), and that was the base level of the chassis done.

▶ I removed everything from the chassis and gave it all a nice tidy up with a grinder and orbital sander. This is, after all, the foundation of my special. *(Chris Hill)*

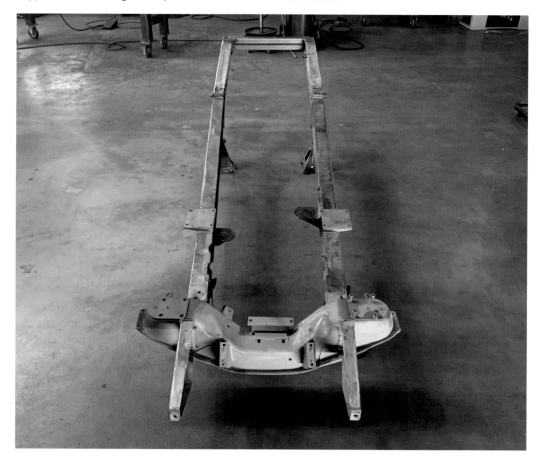

CHAPTER 11

Adding a Panhard rod to the axle

The point of planning the rear axle geometry was to focus on three main axle movements:

- I **DON'T** want my axle moving backwards and forwards. I want the wheelbase of the car to be constant. The strong, fixed, steel trailing arms I have fitted now prevent the axle from moving backwards and forwards.
- I **DO** want my axle to rise up and down over the bumps in the road. The fixed trailing arms now provide the pivot point to allow this, and I will later add coil-over-dampers to complement them.
- I **DON'T** want my axle to move left and right, and at the moment the trailing arms, and even adding the dampers won't prevent that. So I need to add something else.

Panhard rod v Watt's linkage

There are two mainstream ways to restrict the lateral movement of a live axle – a Watt's linkage or a Panhard rod. I often get asked the question, 'Which is better?' The answer is – it depends.

The Watt's linkage involves two bars – one on each side of the car, joined by a single vertical link that attaches to the diff casing via a central pivot. This mechanism ensures the axle stays in the same lateral position as it rides up and down on the springs (ie, it stops the axle from moving side to side).

The Panhard rod is usually a single diagonal or straight bar, one end attached to one of the axle tubes, the other to the chassis.

Supporters of the Watt's link say the system provides identical handling characteristics in left or right corners. Fans of the Panhard rod say it's superior because it's light and simple. The truth is, both systems have their merits.

As the suspension compresses, a diagonal Panhard rod does slightly force the axle off-line. This can add a weird rear-steering element to the geometry.

In a vehicle with a lot of suspension travel, like an off-road truck for instance, there will be noticeable differences between the handling characteristics of a Panhard rod and a Watt's link set-up. However, my car only has around 2in of suspension travel, so the

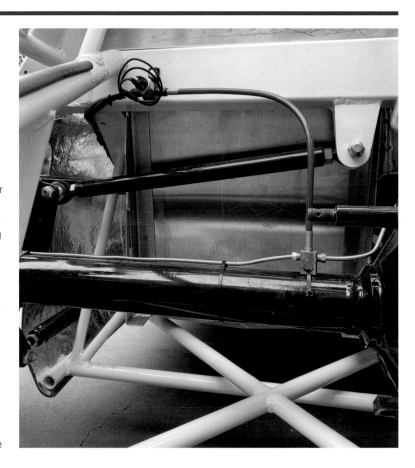

▼ **A Watts linkage eliminates the odd 'rear steering' effect of a Panhard rod, but is more complicated, with a larger number of moving parts.** *(Paul Cameron)*

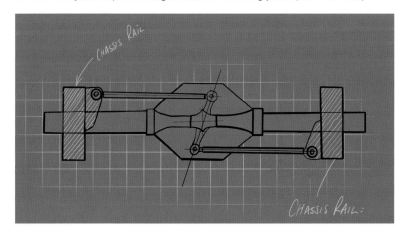

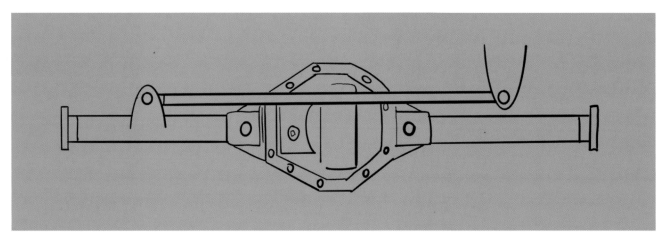

▲ **This drawing shows how simple a Panhard rod is. This is all that's needed to stop the axle moving left and right.**
(Paul Cameron)

difference between the two systems is tiny. For this build I chose the simpler Panhard rod.

Panhard rod

Using the same 1¼in steel tube that I used for the trailing arms, I cut a length 13¾in long. I inserted the remaining two mushroom inserts (see page 59) and

tack-welded them in place. I then fitted the locking nuts and the rose joints, winding the rose joints into the threads so that the eye-to-eye measurement along the rod was 18in.

Now I need to do some explaining! I'm the type of chap who can't switch off. So I go to bed and obsess about the details, have an idea in the night,

▶ **With the Panhard rod and rear axle resting in place, she's starting to look like something you might recognise as a car. Slowly but surely...**
(Chris Hill)

then wake up and simply execute that idea. It's an issue I have suffered from for years. I recall some years ago designing a car from scratch, and after spending around a month building, shaping and contouring the rear end of the body (a staggering amount of work) I came in one morning, cut the entire rear end off and started again from scratch. And that's what I did with the Panhard rod set-up.

Originally, I used a longer, 28in Panhard rod, which you can see in some of the photographs throughout this book, and I secured one end to the threaded insert fitted previously to the front of the axle tube (see Chapter 8), and the other end to brackets on the chassis. I later decided that I wanted the Panhard-rod mounting on the axle to be higher than I had originally designed it, so using two pieces of 4mm mild-steel plate, I made a new bracket which bolted to the original Panhard-rod threaded insert on the axle tube and to the upper trailing-arm mount above it.

On the underside of the new chassis rear crossmember (welded in place in Chapter 10), I mounted a pair of 4mm mild-steel brackets, bolted (using a ½in nut and bolt) either side of a ½in rose joint. I then welded the brackets in place on the crossmember, and fitted the new, shorter Panhard rod between the new axle bracket and the crossmember brackets.

The axle was now firmly located, and yet it could still move up and down. Simple, and a genius piece of engineering by Parisian engineer René Panhard, who pioneered the Panhard rod, among other automotive-engineering feats!

I tackled the final part of the rear suspension (the coil-over-dampers) later, as I wanted to patiently wait until the body was built, to ensure that the springs

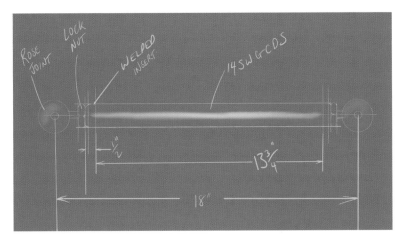

and dampers fitted within the tail section of the body and complemented the look of the car.

In the meantime, I had the remaining two of the three big chunks I salvaged from the donor Alfa Romeo Spider sitting on the floor and staring at me. The next stage was to move on to mounting the engine and gearbox.

▲ The dimensions of my final, shorter Panhard rod (see text).
(Paul Cameron)

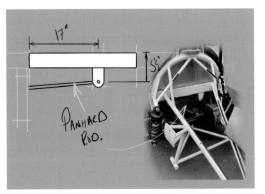

◀ The 'MkII' version of my Panhard rod bracket. Don't worry about making changes like this. If it isn't right, change it – there's no point in convincing yourself something is right, when it isn't.
(Paul Cameron)

◀ These photos show the new two-piece bracket I made for the 'Mk II' higher Panhard rod mounting on the axle. The bracket fits between the threaded insert on the axle tube and the top trailing arm mounting. The last photo shows the Panhard rod attached to the top of the bracket with a bolt and nut. *(Ant)*

CHAPTER 12

Installing the engine and gearbox

I previously removed the engine and gearbox from the Alfa Spider in one big lump. Both would be the subject of a full rebuild, but this was as good a time as any to mount them in the chassis, and work out all the new engineering. First, though, I needed to remove some parts.

I stripped the engine down to the bare block and head, even removing the sump (the Spider sump is winged and fouled the MG TD chassis rails). I didn't touch any of the internals, and I kept the gearbox mounted on the engine to assist my mock-up.

Engine mountings

The engine-mounting brackets on the chassis are really simple. Basically, I created a 'shelf' (or 'platform') on the inside of each chassis rail, and used steel mounting brackets on the engine block. Between these two I inserted rubber engine mounts from a Land Rover Defender, Discovery or Range Rover – really easy to get hold of, universal and perfect for the job at 3in diameter.

With the engine supported on the engine crane, and the gearbox supported underneath with a jack (or an axle stand), I then ensured the assembly rested in the correct position within the chassis, and double-checked that there was sufficient clearance around it. Remember that the engine will vibrate and also twist slightly under load, so I needed to allow clearance for the movement of the clutch slave cylinder, as it sat tight to the right-hand chassis rail.

▶ These engine mounts are very common. The Land Rover part number is KKB103120. I keep a load of these on a shelf, and they get used for all sorts of things. (Chris Hill)

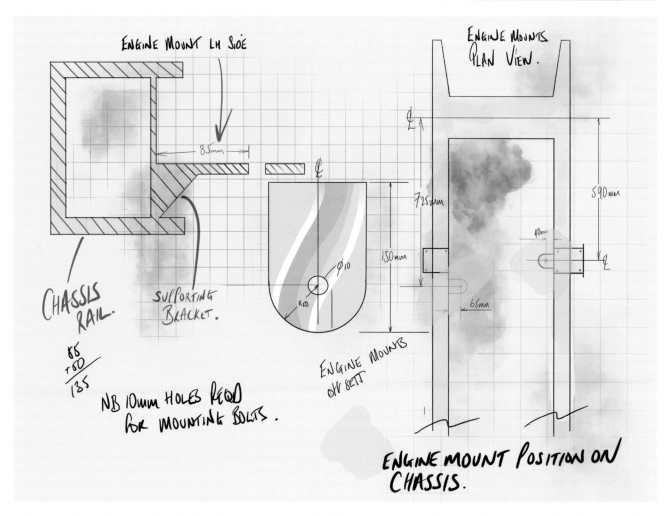

ENGINE MOUNT LH SIDE

ENGINE MOUNTS PLAN VIEW.

85mm

CHASSIS RAIL.

SUPPORTING BRACKET.

Ø10

RªD.

725mm

590mm

130mm

40mm

65mm

ENGINE MOUNTS OFF SETT

85
+60
135

NB 10mm HOLES REQD FOR MOUNTING BOLTS.

ENGINE MOUNT POSITION ON CHASSIS.

To make the engine mountings, first I cut two 'platforms' from 4mm mild steel, and drilled a 10mm hole to accommodate the rubber 'cotton reel' mounts.

I then tack-welded the 'platforms' at 90° to the centreline of the inner side of the MG TD chassis,

and added triangular gussets on the underside to secure and strengthen them.

Using the old Alfa Spider engine-mounting brackets from the block as a template to ensure the correct positions for the holes to bolt the new brackets to the engine, I created two 4mm mild-steel

▲ **Details of the engine mountings. I used 4mm steel plate for the mountings, but 6mm plate could be used for extra strength.** *(Paul Cameron)*

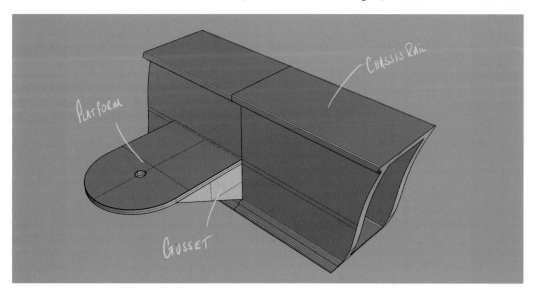

CHASSIS RAIL

PLATFORM

GUSSET

◀ **Make cardboard templates for the components before committing to steel. This makes it much easier to rectify mistakes before committing to cutting metal.** *(Paul Cameron)*

▶ **The engine-mount dimensions may differ depending on the specific circumstances. The components should be tack-welded together to begin with. There will be a chance to go over the whole chassis with the welder at a later date.** *(Paul Cameron)*

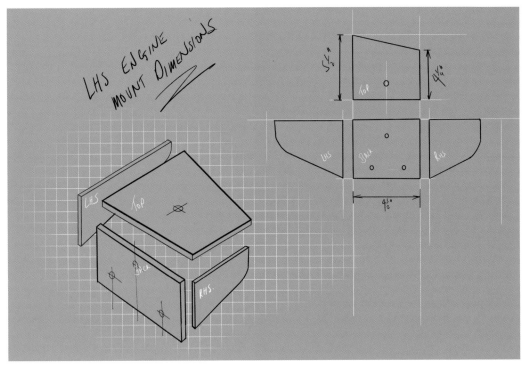

flat plates to replace them, and bolted them securely to the block.

I then double-checked again that the engine and gearbox were correctly positioned and resting at the correct height. With the winged Alfa sump removed, it all fitted nicely between the chassis rails with plenty of space to spare, but with the sump removed I had to be extra careful with the exposed engine internals!

Even with the engine in place, I had a nice amount of space to work around it. I placed the rubber mounts on the chassis platforms fitted previously, then simply created a top platform on each side of the engine, level with the top of the flat engine plate (bolted to the block previously) that covered the entire rubber mount. I first made cardboard templates for these (with a hole in the correct position for the rubber-mount stud to pass through), then transferred to 4mm mild steel.

I then fitted the steel top platforms to the rubbers and tightened the nuts. The top platforms were now held nicely in place by the rubber mounts, so I tack-welded the top platforms to the tops of the flat plates mounted on the engine. I added side gussets for strength (allowing sufficient space around the

▼ **The individual shapes of the engine-mounting components might look very odd, but will make sense when they are assembled.** *(Paul Cameron)*

▶ **I finished off the brackets with gloss black powder coat.** *(Chris Hill)*

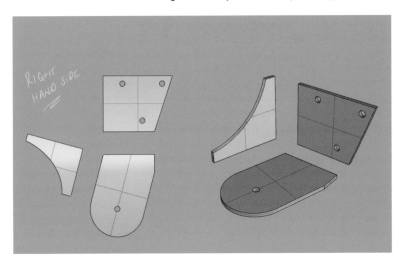

flexible rubbers to ensure that no metal parts made contact with them). Note that I had to shape the right-hand brackets to clear the starter motor.

I made strong and frequent tack welds to securely hold the components together, then removed them from the engine and fully welded all the engine-mounting-bracket components.

Next, I refitted the finished engine mountings to the block, and with the rubbers bolted to the chassis brackets, securely tightened all the fixings. After rechecking that all was correct and level (and making small adjustments using the jack), I moved on to the gearbox mounting.

Gearbox mounting

It was far easier to tackle the gearbox mounting using the actual gearbox as the jig. With the engine already secured using the new mountings, the whole assembly was held firmly in place.

The existing Alfa rubber gearbox mount is actually pretty weak. I even tried installing a brand new one, but to no avail – it was just way too flexible for my needs. I decided to remove it and make my own mount.

Using a piece of nylon 'bar' (easy to source from numerous stockists) I turned two mushroom inserts on the lathe. nylon is a joy to work with on the lathe – it's easy to machine.

I gently tapped the finished new nylon inserts into the gearbox casting – they were a nice snug fit! This created a firm mounting point to work from.

Using the 1in solid-mild-steel bar (the same bar used for the trailing-arm front mountings on the

▲ With the engine/gearbox assembly lowered into place on the engine mountings, I could then tackle the gearbox mounting. *(Ant)*

◄ I used nylon to make a new gearbox mounting, which was more rigid and durable than the original Alfa Spider rubber mount. *(Ant)*

◄ Nylon is a joy to work with on a lathe. It's an idea to go slowly and take regular measurements, as its easy to remove too much material when the job is so much fun! *(Ant)*

▲ **The same 1in steel bar used for the trailing arm front mountings came in handy for the gearbox mount.** *(Ant)*

chassis), I cut a piece 5in long. Using the lathe, I then drilled a 16mm hole through the length of the bar. I chose to use 16mm, as that's the diameter of the original Alfa gearbox-mounting bolt. It's possible, of course, to machine the nylon to suit any bolt, but it needs to be a substantial one – anything less than ½in (M12) would worry me.

I then made a 3in-diameter washer from 3mm mild steel (as an off-the-shelf 3in washer with a 16mm hole was hard to find) and placed it against the nylon insert. Using a 9½-in-long 16mm bolt, I then passed the bolt through the steel sleeve and washer, through the nylon bush, and secured in place with a nut on the other side of the gearbox. The assembly was nice and rigid.

Due to the angled casting of the Alfa gearbox, the steel sleeve naturally wanted to tilt slightly down. By

using the gearbox as a jig, the sleeve was now held in the correct position, and that was not at 90° to the chassis rail.

Next, I cut two pieces of 3in, 4mm mild-steel flat bar (the same material I used to triangulate the front lower trailing-arm mounting on the chassis). Just as I did with the lower trailing-arm mount, I then used the two pieces of steel bar to triangulate to the gearbox-mounting sleeve. I added a third, vertical piece of steel, at 90° to the washer (see photographs below) and welded it all up. I welded the washer in place, as this created a surface to spread the load. I carefully tack-welded it all first, taking care not to apply too much heat to the washer, which might have melted the nylon inserts. Once tacked, I moved the gearbox away, then fully welded.

That was it! The engine and gearbox were fully secured in place.

The build was in a bit of a 'Catch 22' situation at this point. The mounting of the engine was complete, and it was best to complete the remaining chassis work with the engine in place, but I needed the body fitted to my chassis to complete the chassis work.

So, to enable me to keep making progress with my build, I removed the engine and gearbox at this stage and sent it straight to the machine shop for strip-down and a full check.

I simply exchanged the gearbox for a reconditioned unit I found online, as the exchange 'box was ready to go and affordable. The fact that I exchanged the old 'box saved a few quid too. I could have just cleaned up and reused the original gearbox, but I opted to use a reconditioned 'box, keeping me on track for my car, my way.

▶ **With the gearbox held in the correct position, I could custom-build the mount to suit.** *(Ant)*

▶▶ **Note I ran the vertical gusset right to the top of the chassis rail, utilising all the strength of the entire chassis box section. I also removed the rubber 'donut' from the gearbox output flange.** *(Ant)*

◄ **A fully reconditioned Spider gearbox in exchange for my tatty old one. I'll take that.** *(Ant)*

I also sourced a new slave cylinder, clutch and a lightened flywheel.

Unlike the gearbox, a lot of time was spent rebuilding and upgrading my engine.

Engine rebuild

With a project using an engine with unknown history, there are a number of options. The donor engine could be used in as close to 'stock' form as possible, maintaining its heritage. Alternatively, the components could be lightened and balanced, and maybe a turbo or a supercharger could be added. Perhaps even nitrous oxide could be used to take the load on the engine components to bursting point!

Whichever direction is taken, the process will start with a strip-down. The strip-down will reveal a lot about the engine's history, and might dictate how the rebuild is tackled. It's worth documenting as much detail as possible, and taking close-up photos of any scoring, or wear marks visible. A specialist engine shop will measure critical points for wear, like the cylinder-bore diameters for instance, and can advise on what needs replacing.

After careful diagnosis my engine needed:

- Crankshaft balancing
- New cylinder liners
- New pistons and rings
- Connecting rods machined to fit the pistons

I also chose to gasflow the head. This involves smoothing and sometimes reshaping the inlet and exhaust ports and combustion chambers – from the Weber carburettors, the mounting adaptors, through the intake manifold and into the cylinder-head ports, so they all match perfectly (also called 'port matching'). This process results in a really smooth journey for the air and fuel, from outside the engine to the combustion chamber. It's slow and laborious work, but done by a skilled expert can make all the difference to engine power, as the flow of fresh air/fuel mixture is vastly improved. I chose to add larger valves too, increasing the quantity of air and fuel entering the combustion chambers.

While the engine was receiving a new lease of life, I cracked on with my chassis work, but to do that, I needed a bodyshell!

THE ALFA ROMEO 158 ENGINE – THE HEART OF THE MASTERPIECE

▶ **The 158 engine in all its glory.** *(Centro Documentazione Alfa Romeo – Arese)*

I love the look of the later 158, and that's one of the main reasons I wanted to do this project, but the heart of the real 158 was its utterly remarkable engine. For a start, just contemplate the thinking; designing a whole new straight-eight engine for a car that was only ever going to be a 1.5-litre supercharged Voiturette Class racer is, at first glance, puzzling, until you realise that Alfa was also developing a supercharged 3-litre V16 (the Tipo 316) to take on the might of the German Silver Arrows in full-fat GP racing. The two projects, while not interchangeable as such, definitely shared intellectual effort at the very least. Although the engine was designed by the overall father of the 158 project, Gioacchino Colombo, it undoubtedly owed much to his mentor, Vittorio Jano, (who left Alfa after the perceived failure of his 12C at the end of 1937, moved to Lancia and ultimately Ferrari), so much so that when Jano was looking at Colombo's drawings he apparently exclaimed, "It's one of my engines!" Colombo had undoubtedly learnt from the master Jano, who he'd been working under since 1924. However, by late 1937 he was developing his own ideas, working at Ferrari's works 200km (124 miles) from Alfa's factory in Portello, and had lost Jano as a regular collaborator. After Colombo died, papers revealed he had corresponded in detail with British engineer (later Sir) Harry Ricardo, who had a significant influence on the detail engineering of the 158 engine.

The 158's straight-eight twin-overhead-camshaft engine featured a finned cast Electron crankcase split along the centreline. The crankshaft ran in seven main bearings with an eighth outrigger bearing adjacent to the flywheel. The crank was machined from a billet of chrome steel. The aluminium cylinders comprised two castings bolted together, and dry cylinder liners were used. A single spark plug per cylinder was fired by two two Marelli magnetos driven from the front of the engine. There was one big difference between the V16 and the straight-eight. In the V16, the geartrain driving the camshafts and ancillaries was at the rear, and in the straight-eight it was at the front, allowing the crankshaft to be shorter. The 158 engine was initially fitted with a single Rootes-type supercharger, but as the engine evolved, so did the supercharger, and by 1940 a two-stage supercharging system had been developed. Each cylinder of the 'eight' was just under 185cc, but contrary to later Alfa engines, it was under-square with a bore of 58mm and a stroke of 70mm. That should restrict its revs, yet by the end of its racing life, in 1951, it was revving to 9,300rpm and producing nearly 430bhp – awesome! With a fuel consumption of 1.4mpg or less, the 158 burnt well over 200 gallons of fuel in a GP, which was ultimately its undoing. That's quite a development achievement when you consider that in its first race it was rated at 195bhp@7,000rpm – a more than respectable output for a supercharged Voiturette in 1938.

Amazingly, Alfa only ever cast nine 158 cylinder blocks, in two batches, and even when they were quite badly damaged, they repaired them to save money. The near-430bhp engines Fangio used to win the 1951 World Championship (in the 159) used those same 14-year-old castings! They had a weak point though, as the main-bearing caps were held in the magnesium crankcase by two short bolts, and above them were short hold-down bolts for the cylinder blocks. Between the ends of these upper and lower bolts each of the original four crankcases developed large and serious cracks. Colombo pleaded for new castings in which he

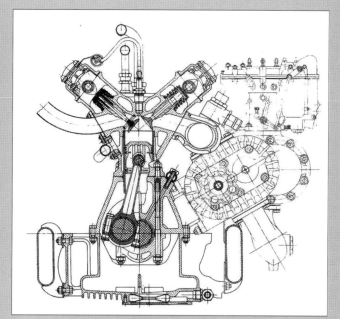

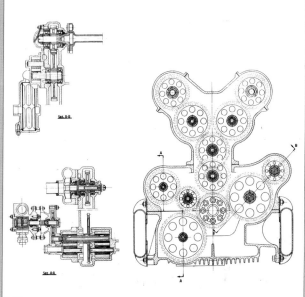

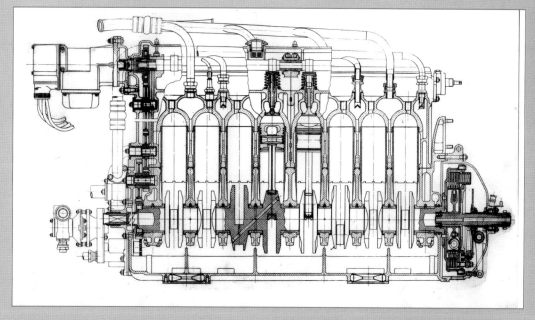

Three cutaways through the 159 engine from front, rear and side. *(Centro Documentazione Alfa Romeo – Arese)*

could run big single bolts from the bearing cap to the block to eliminate this, but this suggestion was refused on grounds of cost. So, he drilled the cracked cases and installed new through-bolts, and in this patched-up form the engine went on to develop ever-higher outputs for another dozen years. At the time, this modification was kept secret.

The other problem was a big-end-bearing weakness, and this was rectified for the 1939 season with the adoption of needle-roller bearings, which were made in two microscopically different sizes, one being 0.0008in smaller than the other.

Once those two issues had been sorted out, Alfa kept blowing more supercharger pressure into the engines to get more power, and they never complained. Eventually, the two-stage blower was running at 42.6psi (that is not a typo!) and using fuel that was 98 per cent methanol, for combustion-chamber cooling. For the brave, that gave the Alfetta a maximum speed of over 190mph, and at tracks like Reims, which was basically a big triangle, that speed would have been maintained for quite long periods of time on skinny tyres in a car with no seat belts or roll cage. Fangio won there in 1951, completing the 374-mile race at an *average* speed of 111mph. The engineering may have been magnificent, but the men who drove the finished car were beyond brave.

CHAPTER 13

Building the body buck

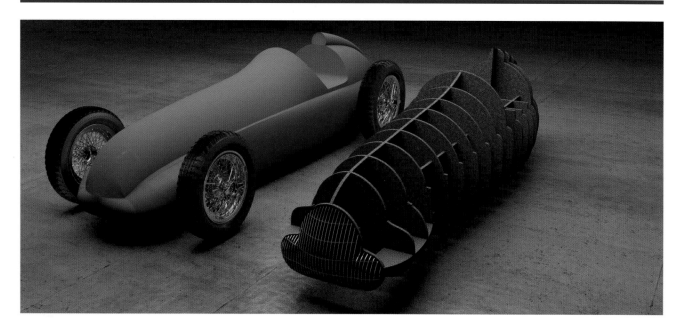

Getting your head around the bodyshell is a HUGE part of the build – this is how the car will be seen by the world, and there are numerous options available.

Although I could have chosen to shroud the car in any shape I wished, I chose a boat-tailed-style grand prix car. That's the whole beauty of building a special – there are no rules. There is a world of inspiration out there.

I simply drew inspiration from the Alfa 158 – I never set out to build a tool-room copy or exact replica. For me, that is not what special building is about. I tend to think of it as more of a homage –

taking design nods from the original, then adding my own twist.

One thing experience has taught me is that the key to car design is proportion. The car needs to sit well at all angles. If one element is too long, too short, too high or too low, it throws the look of the whole car into disarray. The Alfa 158 is just right – she is beautiful from every angle, so proportion-wise I changed nothing. My car would be faithful to the original in terms of length, height and width, and as we already know the wheelbase and track are accurate. Sticking to these known

▶ The essence of good car design is that it must look good from all angles. The Alfa 158 does just that.

(Paul Cameron)

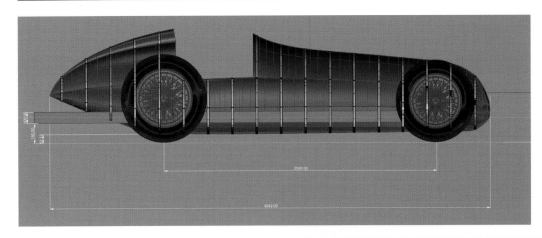

quantities ensured that the finished car would have the look and feel of the 158, and the chassis would eventually fit the body.

As we know, the car sector has evolved over the years, and it's still evolving. Today's modern Formula 1 cars use state-of-the-art carbon composites – lightweight and strong. Composites were not an available option to Alfa Romeo in the 1930s when the 158 was designed and built. Had they been, I am convinced that they would have been used.

However, in the 1940s, the car world changed forever with the introduction of glassfibre. For many reasons, my choice was to make a one-piece bodyshell completely from glassfibre. To achieve this, I needed to create a mould, and to make a mould, I first needed to create a 'buck'.

If I had chosen to take a more traditional route, and hand-form an aluminium shell, I would still have needed a buck over which to form an accurate shape. Either way, a buck is next.

Turning again to my friend Paul, he created the CAD (Computer Aided Design) data to capture my desired boat-tailed, single-seater 158 shape in 3D. The CAD data enables accurate dimensions to be determined.

The great thing about CAD is that it has opened up a whole range of options for car design and building. Already CAD had helped me to create the data to water-cut steel parts for my rear suspension, and now it would help with my body.

Drawing anything using CAD requires plenty of reference material, and it's important to have as many different elevations (views) as possible to hand. As the car is drawn, it's possible to check that the lines work in different orientations. The idea is to build a 'surface' of the body you want. Once you're happy with the overall shape, the computer can be used to slice the model up into 200mm sections. These sections became the ribs of the buck. The CAD approach allows the finished project to be visualised, while also generating the CAM (Computer Aided Manufacturing) tool paths needed to machine the buck components. It's worth bearing

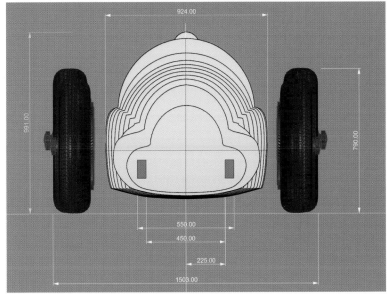

in mind that during the CAD process it's possible to incorporate location points where the chassis should meet the buck for alignment purposes. The CAD approach saves a lot of time and money, as any changes can be made before committing to cutting the material.

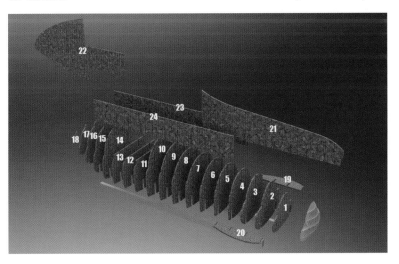

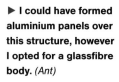

▲ **This is a really fun process. It reminds me of those model dinosaurs I used to make as a (smaller) kid.** *(Ant)*

Taking Paul's CAD data, we started by splitting the car into cross-sections. I was then able to transfer that data to a CAD milling machine and have each section cut for me from ½in-thick OSB board, and numbered.

The beauty of CAD is that it can also be used to alter the scale of a component, making a shape bigger, smaller, wider or taller. I stayed quite close to the original 158 size and dimensions that I was trying to capture, but I made a few subtle changes.

Organising the OSB boards into the correct sequence was essential. The CAD design made provision for slotting the parts together, and it was an incredibly accurate and very satisfying process. It reminded me of those click-together kids' wooden dinosaur models.

The whole assembly slotted together in minutes, and it also formed a self-supporting structure, holding itself firm, though I did add some wood screws to some of the joints for extra rigidity.

If I was building the bodyshell in aluminium, I would still have required the buck as far as this stage, but at this point I would have used it as a wooden former to shape the aluminium panels over.

However, I opted for a glassfibre body, so I needed to do a great deal of work to turn the OSB formers into a solid buck that could be used to make the mould. I started by filling in the gaps.

I wedged 4in foam insulation board (which is used in house builds) into the gaps between the formers, and fixed it with a thin squeeze of wood glue. Once I'd filled in all the gaps, and the glue was dry, I then shaped the foam with a bread knife (yes, a bread knife!), which I found was really great for creating the contours of the body shape.

I covered the joints between the boards with silver tape, so the shape flowed smoothly.

▶ **I could have formed aluminium panels over this structure, however I opted for a glassfibre body.** *(Ant)*

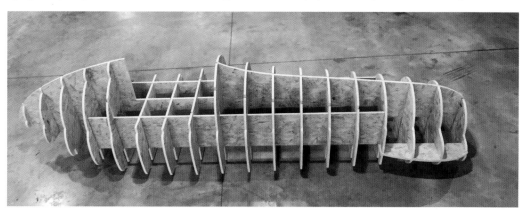

▼ **This photo shows the front end fully filled in with foam and sealed with glassfibre, and the middle and rear sections just foam.** *(Ant)*

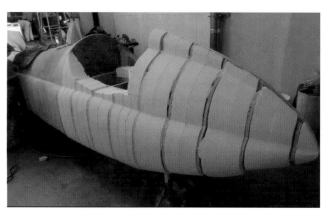

▼ **Adhesive foil tape is brilliant at covering the joins. It's starting to take shape.** *(Ant)*

Once the shape was set using foam and tape, I applied a thin layer of glassfibre over the top, then spent hours and hours – no, wait – days and days, gradually applying layers of body filler, and sanding down to get a perfect, smooth shape, with a few helping hands from good friends.

I used over 30 large tubs of filler (I also used stopper and glassfibre resin) for the initial layers, applying thin layer after thin layer, sanding each layer down before the application of the next skim. This was repeated until the svelte shape was perfect. Carrying out this procedure may work off a few pounds, but the labour is worth it when it comes to creating the mould. Any snag, or even the smallest scratch or imperfection, will be reproduced on the surface of the mould, and will be captured forever, so it's extremely important to get this right.

Smoothing out a large surface is best tackled with a flat-bed sander, and any curves or swage lines can be tackled by hand using a rubbing block. It's essential the shape flows from nose to tail – do not use the flat of your hand for sanding flat areas, as the pressure points from your fingers will cause a 'finger ridge' or furrow effect where the surface becomes uneven. As explained previously, the best way to get a perfect shape is by applying thin layers of filler, leaving each to harden, then

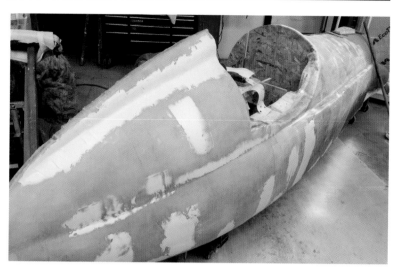

sanding down, and repeating the process until the shape is perfect.

I started with a heavy-grit abrasive paper to get the rough bulk of the shape, using progressively less-coarse paper as I went. It took many, many hours, but it was worth investing the time at this stage. I started with P80-grade abrasive paper to tackle the fresh filler quickly, then I used less-coarse P180 on the final skims, finishing with P320 (the higher the number the finer the grit).

▲ It takes many days to achieve the perfect finished shape, but it's very much worth it. *(Ant)*

▼ Many, many thin layers of body filler, applied gradually and sanded to provide a smooth shape. I used over 30 large tubs of filler! *(Ant)*

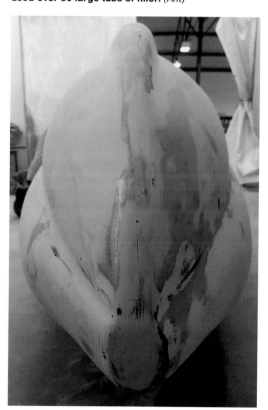

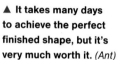

◄ A flat sander like this is perfect. This is an air tool. Battery sanders are very good these days and, of course, are cordless. *(Mac Tools)*

◄ All hands on deck! My pal Matt Lawson helping to get the body into shape. *(Ant)*

Once I was happy with the surface, I removed surface dust, then wiped the whole buck with a degreasing agent. It was important to wipe it dry to avoid any chalky residue – in other words, wipe on, wipe off. I find it also helps to inspect the surface by adding this wet sheen of degreaser, which shows up any ripples or defects when looking down the surface. Using degreaser is also important as part of prepping in advance of applying primer, as it will remove any dirt or contaminants, ensuring that the primer adheres correctly.

It was then time to move the buck into a paint booth and load it up with some heavy, but even, coats of high-build filler/primer. This stage is essential, as the primer will fill in any scratches, pinholes or defects. It's important to follow all the safety recommendations for the specific product being used when spraying – especially protective gear. It's also important to allow adequate flash-off times (the intervals left between coats) – normally 10–15 minutes, depending on the product data sheet for the type of primer being used. Leaving the correct time between coats will allow solvents to escape, avoiding possible microblistering from trapped solvent. I usually apply three or four coats of

▶ **Adding a heavy coat of primer fills any tiny scratches and gaps. Be sure to read the instructions on drying times and PPE.** *(Ant)*

▶ **It's important not to bake the buck! The buck should be allowed to air dry, because trapped air within the glassfibre or body filler can heat up and expand to potentially crack the finish or cause air bubbles.** *(Ant)*

high-build primer, as this gives plenty to work with to achieve that perfect finish when it is dry.

Note: *When working with isocyanate 2-pack paint, a full air-fed mask and all the associated recommended personal protective equipment* **must** *be worn. We opted for water-based paint products.*

Having allowed the primer time to cure, I needed to flat the primed surface to prepare it for the moulding stages. It was crucial to polish out any imperfections in the surface and to help with this, I applied a 'guide coat'. This normally comes in a black-powder form, however a matt-black aerosol works just as well, and a light dusting over the entire buck will highlight high and low spots, scratches and ripples, which can then be rectified at this stage, leaving a smooth surface.

I used a rubber flatting block, some P800, P1200 and P2000 silicone-carbide wet-and-dry, and a bucket of water, keeping a strong arm at the ready. I usually avoid trying to cut corners by using a DA (Dual Action) sander, as these can be too aggressive and can take away more primer than required, leaving a wavy surface, which may not be visible at this stage, but will stand out like a sore thumb as soon as the shiny coat of paint goes on the bodyshell. Using a flatting block avoids this, and helps to make sure the surface is flat. It was worth spending the extra time to get it right! It would be worth it in the long run.

At this point, I had flatted the entire buck with P800 wet-and-dry, and rectified any defects, because the buck needed to be super smooth. The buck then required flatting with P1200 paper, then P2000, to bring the surface to a near-polished state. I repeated the process again, using a guide coat for the P1200 and again for the P2000, and I then dried the surface and removed all the residue.

The whole buck required a polished finish – the more polished the surface, the better the mould would be. A buck should look and feel like a sheet of glass, and to achieve this requires proper polishing equipment. I used a polishing mop (electric rotating polisher) and three grades of flatting compound – coarse, medium and fine. I used three different mop heads to match each of the compounds, starting with coarse compound, and keeping the surface clean. I repeated the procedure through each stage until I achieved a shine, then I wiped everything clean, at which point it was time to put the kettle on!

It was a great deal of work to get to this stage, but that was only one third of the way to making the final bodyshell. So far, I had only created the buck (I eventually threw it away once I had made the mould – it was amazing how heavy the buck was when finished!).

The next stage was to create the body mould from the buck.

CHAPTER 14

Building the body mould

Mould making is pretty ambitious. It is a specialist skill and incredibly easy to get wrong. I have spent years ruining moulds and making many, many mistakes. It requires a good eye for detail, a huge amount of patience, and the key to getting it right is experience.

The best way to describe the 158 body mould is that it is like making a chocolate Easter egg. You need to be able to remove the egg from the mould, so because of its shape, you can create a single mould to make half an egg, but not a whole egg. To make a whole egg, you need a mould in two halves that join together, and that then come apart to release the egg.

The good news is that the shape of the 158 is incredibly simple – not as simple as a chocolate Easter egg, but imagine creating moulds for today's complicated racing cars. The 158 shape is actually a good one to cut your teeth on.

First, I needed to ensure that the finished, primed surface of the buck was as smooth as possible, with no flaws, scratches or imperfections. Even a human hair on the surface would translate into the mould and remain forever.

The shape of the 158 lends itself perfectly to a centre-release mould (the same as the Easter egg), so I planned to create one half of the mould at a time.

To make the first half of the mould, firstly I placed a masking tape line along the top of the buck to act as a centreline (ie, one edge of the tape was run along the top centreline of the buck). This was the reference line I would be working to. Against this line (the edge of the tape) I placed a strip of 4mm acrylic board – I chose the type of board used for advertising boards (think a generic estate agent's 'For Sale' board). This material is easy to cut with scissors, and has a shiny, glossy surface that's easy to release from when moulding the body. I cut strips of the board about 3in tall to follow the shape and contours of the body. I used a hot-glue gun to glue the strips to the edge of the masking tape, and then used the glue gun again to stick some small wooden tabs in place, at an angle, to help reinforce and strengthen it all. I repeated this process down the backbone of the entire buck.

On the working side of the acrylic (the opposite side to which I fitted the strips of board), I used my fingers to squeeze some Plasticine into the edges of the acrylic where it contacted the buck, to ensure

◄ **Confectionary and car building are more closely linked than you might at first think.**
(Getty Images)

▶ Creating a foam-core low 'wall', using acrylic board along the length of the body. This begins the process of making two mould halves. *(Ant)*

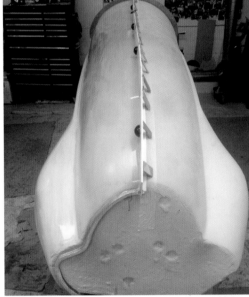

▶▼ You can see the shape of the first half of the mould starting to come together. Note the border all around the mould. The blue sheen is a water-based 'release agent' I hand-painted on with a paint brush. *(Ant)*

an airtight seal at that lower edge. I found that preheating the Plasticine with a heat gun in advance really helped. I then scraped away the excess Plasticine with a sharp knife, so that all the gaps were filled and the edge remained sharp, with a 90° angle between the buck and acrylic board.

Next, I got hold of some squash balls (yes squash balls!) and cut them in half on a band saw (a sharp blade would do just as well). I wanted these to create domed shapes within the edge of the mould that would later act as location 'pegs' to lock the mould halves together (to ensure correct alignment).

I used the glue gun again to stick the half balls to the edge of the board every 20in or so, and again I used Plasticine to seal around the tiny imperfections in the edges of the balls. Note I also covered the joints between the sections of hand-cut acrylic sheet (the vertical joints between each section) with silver-foil tape. Essentially, I was creating the return edge of my egg mould.

I also built a similar return edge on the lower part of the body, and again used strips of shaped acrylic, silver foil tape and a series of half squash balls. This lower edge was just for safety, to make the return for the underside of my body easier to release. It probably wasn't essential with the easy, smooth shape of the 158.

I made sure I created a 90° border around one entire half of the body buck (front and back as well) and checked that the small gaps were filled with Plasticine, and that the joins were covered in foil tape.

After finally checking that the surface of the buck was smooth and free from contamination, I covered it with a smooth, even coating of water-based release agent. The release agent creates a thin barrier between the actual mould and the fibreglass

▼ The release agent must be applied evenly all over the surface. Many, many hours are required for this stage of the build. *(Ant)*

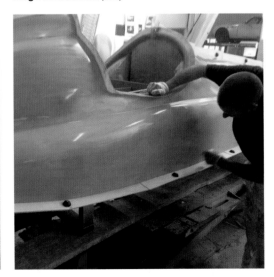

GEL COAT

The gel coat is like the outer shell of an egg – it is brittle and very hard. On its own, gel coat is pretty useless, and it needs the strong glassfibre layers behind it, but it is the surface that you instantly see and touch, and the surface that actually gets painted. Some car manufacturers actually add colour pigments to the gel coat to provide the finished surface, which does not then require painting.

Like much of the mould-making process, to produce a gel coat, materials are mixed to start a chemical reaction. The gel coat itself is a polyester resin, which is a completely synthetic material. The resin is a liquid that will stay liquid until a catalyst is added to harden it – the catalyst starts a chemical process that can't be reversed. The usual catalyst for polyester resin is methyl ethyl ketone peroxide, or MEKP, which is nasty, poisonous stuff. It's necessary to wear gloves and take suitable safety precautions when handling these products.

I mixed the gel coat and the catalyst together, following the manufacturer's recommendations to make sure the ratio was correct. I had to ensure that the two products were well mixed – the last thing I needed was to find a bit of gel coat that hadn't cured.

layer, to prevent it from sticking – similar to the non-stick coating on a frying pan.

On top of the release agent, I brushed a liberal, but even, layer of gel coat. I took my time with this process, as this layer of gel coat would be the surface used to create the body.

I brushed the gel coat on to the buck as evenly as possible, ensuring that any bubbles were removed as brushing progressed. At this point the gel coat was already starting to cure in the container, so there wasn't a vast amount of time, but I also had to avoid the temptation to rush. This is what makes these processes fun for me – it's all about managing time and keeping your cool – one mistake can easily set you back hours or days.

I continued to brush gel coat on to the buck until it was all covered. I made sure the gel coat covered the surface evenly, but not too thick, as otherwise micro-cracks can appear in the gel-coat layer. I had to take my time, but not dawdle – did I say the material was already curing in the pot?

Cure, set, harden, go off – they all mean the same thing, which is that the material transforms from a liquid into a solid. The mixed two components of gel coat need heat to cure – when the two chemicals are mixed, an exothermic reaction starts

to happen, which means that the chemical reaction creates its own heat.

Note: *When working with the gel-coat constituents, it's important not to mix too much material at once. If a whole bucket was mixed in one go, for instance, the heat generated would be huge, and it could cure in a minute or so. I've seen mixing tubs churn out smoke and actually melt due to excess heat!*

If the workshop is too cold, sometimes it may be necessary to bring in some heat lamps to raise the temperature, but it's always important to check the manufacturer's recommendations.

Once that gel coat had set, I then started applying layers of CSM (Chopped Straw Mat) soaked in resin. This adds the strength and substance to the finished layer. Individual flexible glass fibres are laid together to make a mat, with fibres running in all directions. It's similar to the OSB board I used to make the buck. Individually, the pieces are not very strong, but when all laid-up together, and held together with resin, the finished material is very, very tough.

During this process, it was important to ensure that I overlapped the CSM up the walls of the acrylic

▲ **The completed first half of the mould (left of picture) in place on the buck, with release agent on the buck ready to start the second half. Note how the shape of the balls is now transferred into the first side of the mould.** *(Ant)*

▲ **This same technique is used to create all kinds of glassfibre objects, from canoes to industrial vents.** *(Ant)*

board, and also to dab with the brush I was using to apply the resin, so that the CSM followed the shape of the detailed areas within the edges of the mould and around the half balls. This first layer was critical! I also had to take time to ensure that no air was trapped within the CSM and resin.

I built up two layers of 1.5-ounce CSM (this is the weight of the material per square foot) to create a 3-ounce thickness, and once the layers had fully cured, I removed the acrylic strips, the wooden tabs, the half balls and the masking tape. The glue from the hot-glue gun peeled away easily, and that was one half of the mould finished (for now).

I left the first half of the mould in place on the buck, while I built the second half. The acrylic strips were not required for the second half, because I was now going to build the second half up against the fully cured surface of the finished first half.

Before I started the second half of the mould, again I pushed a bead of plasticine into the edge where the buck met the first half of the mould, and I

scraped away the excess with a sharp knife.

Again, I painted release agent onto the buck, making sure I coated the return edge of the completed first half of the mould (still in place on the buck). I then repeat the 'lay-up' process on the second half of the mould, with a layer of gel coat (allowing it to cure), then two layers of CSM soaked in resin.

That was the bulk of the work done. I then built up the two halves of the mould with layer after layer of CSM and resin – the thicker the mould is built, the longer it will last. It's OK to build a lightweight mould if it's only going to be used to make one body, but my thinking was that I may need to carry out a few body repairs in the future, so I built a heavier mould, just in case...

With both halves of the mould finished and cured, I carefully separated them along the top edge, by carefully tapping in wedges, and then lifted each half off the buck. Job done – I now had a mould for my 158 body!

CHAPTER 15

Building the body

I used my facility in the UK for the process of building the body, as the workshop is specially kitted out for glassfibre, with the correct extraction and safety equipment.

Moulding the body successfully is all about the preparation.

With the mould finished as described in the previous chapter, I started by using a cotton cloth to wipe some liquid beeswax all over the surface of the mould. Using a sponge, I then applied a smooth, even layer of water-based release agent, and allowed it to air dry fully.

Getting into the small nooks and crannies with the release agent would really pay off when it came to removing the body from the mould later. Any tools can be used to help to do this – rags, brushes etc – as long as there's an even, thin layer by the end of the process.

The actual process of making the body involves layers – lots of layers – and it starts from the outside surface in. The first layer I laid up in the mould would be the very outside surface of my body – the surface that got painted. The remaining layers would be built up inside that.

I mixed some gel coat (as described in the previous chapter when making the buck), and brushed the gel coat onto the inner surface of the mould until it was fully covered.

▲ Happy in my work. Always read the safety instructions with the products used – some can be harmful to health. *(Master Mechanic)*

▼ Layers must be applied evenly throughout the process, whether it's release agent, gel coat or CSM. *(Master Mechanic)*

▼ The two halves split from the buck. Remember, this isn't the bodyshell, this is the mould to make the bodyshell. *(Ant)*

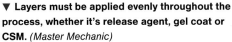

▲ **It's important to work fast but very carefully. It's the precision that draws me to this kind of work.** *(Master Mechanic)*

▶ **Laying up manageable-size sections of CSM, making sure the edges are torn, not cut.** *(Master Mechanic)*

▲ **This is me on television, explaining the roller used to squeeze out air bubbles from the wet glassfibre mat.** *(Master Mechanic)*

▶ **Getting on with it. Removing air bubbles is a full-time job when working with glassfibre.** *(Master Mechanic)*

Making the glassfibre bodyshell

Next, it was time to layer up some chopped straw mat (CSM) – again see the previous chapter for details. I used 1.5-ounce CSM. It can be cut with scissors (but they will blunt quickly), or ripped apart with your hands (gloves should always be worn when working with CSM). I measured each section of CSM and cut it to size before laying it in the mould dry to check the fit.

Any edges that will hang over the edge of the mould can be cut with scissors, but any edges that need to blend with other sections of CSM should be ripped – the ripped fibres allow the CSM to mesh with other sections. If the edges are not well meshed together, it's possible that the finished body could crack months or years later.

Note: *It is necessary to leave a gap before the edge of the mould, where the two halves join together. The aim is to fill as much of each half as possible when the two halves are separate then, when the mould is bolted together, to finish the last joining strip down the middle, joining the two halves together.*

Mixing the resin is very similar to mixing the gel coat. Again, I had to follow the instructions for the ratio, make sure I didn't mix too much at once, and work fast.

I applied the resin on a 'wetting out' board. Ideally, this should be a wooden board or a sheet of plastic that the section of mat lies on while applying the resin with a brush – I used some cardboard. I normally get through a number of brushes for one project – there's no point washing resin brushes out, and they are a cheap, consumable item these days. I usually buy a load of 1in, 2in and 3in brushes.

It's important to apply enough resin to thoroughly soak the CSM through – you can always brush on more resin once the section is laid into the mould. The trick is to get it nice and wet.

I then laid the wetted CSM into the moulds piece by piece, starting with one end, and gently laying it down flat against the gel-coated surface.

With the wet mat laid into the mould, I used a roller to squeeze out any air bubbles. The proper tool is a special grooved roller that looks like a multi-toothed cog. This is the perfect tool as it ensures any pockets of air are rolled out. It is vital to avoid any air bubbles trapped in the material, as they can cause damage to the body in years to come. Even though the resin cures quickly, time must be taken rolling to remove bubbles. The appearance of the mat changes as the resin soaks in – when it goes from white to slightly translucent, it's a good indication that the material is properly soaked.

It's easy to be tempted to lay too much mat at once, but this should be resisted. Smaller, more manageable areas are easier to cope with than fewer large ones. Although the resin cures rapidly, and speed is key, it's important not to mess up.

As soon as one layer of CSM had cured enough that it wasn't sticky, I added a second layer on top of it. In reality, by the time I had got all the way around the mould and back to where I started, the resin had cured enough to start another layer anyway. I kept going until the whole surface was covered by two layers of 1.5-ounce mat to create a 3-ounce-thick skin all over.

I had to take great care, as it was absolutely essential that no material got on the two mould return flanges, or any other joining surface, to avoid headaches later on. If it did, it would get in the way when bolting the mould together, creating a gap.

I then left both halves to fully harden. Once the curing process is underway, it's impossible to reverse, but it's still important to adhere to the recommended temperatures to make sure everything goes to plan.

Next, I bolted the two mould halves together tightly. The half squash balls came into play here, allowing perfect alignment to create my 'egg'.

At this stage, I had a thin strip that needed finishing down the middle. I applied a strip of gel coat around 6–8in wide down the long join area. I overlapped it right over the top of the new glassfibre, making sure I had full coverage. It needed to be as consistent as the main gel-coat layer.

I then repeated the glassfibre procedure with a final strip of resin-soaked CSM down the middle of the join line.

Again, I left the resin to harden. I still had to maintain workshop temperatures and use heat lamps etc, to avoid any issues and make sure everything went to plan.

I left the whole structure for a couple of days, to ensure that the materials cured and that the whole shell would be rigid and tough. Once I was happy that everything had fully cured, it was time to reveal the finished bodyshell.

I used a rubber mallet to drive a plastic wedge between the flanges (formed by the acrylic sheet) where the mould halves were joined. I did this carefully in a few places, releasing all the edges I could get my wedges into, and kept working round the mould. I heard a positive 'pop' as the joint gave and the shell came away from the mould. I breathed a sigh of relief, knowing that it

▲ Even when cured, the material can give you glass splinters, so always wear gloves. (I've forgotten here.)
(Master Mechanic)

◄ Bolting the two halves of the mould together. The cured material is very tough.
(Master Mechanic)

▲ Laying up the CSM evenly and consistently even in the hard to reach spots – that's the challenge.
(Master Mechanic)

◄ Rolling, rolling, rolling. Still working at removing those air bubbles.
(Master Mechanic)

▶ Left to cure for a couple of days, the complete assembly is now quite heavy. Care needs to be taken when moving it around. *(Master Mechanic)*

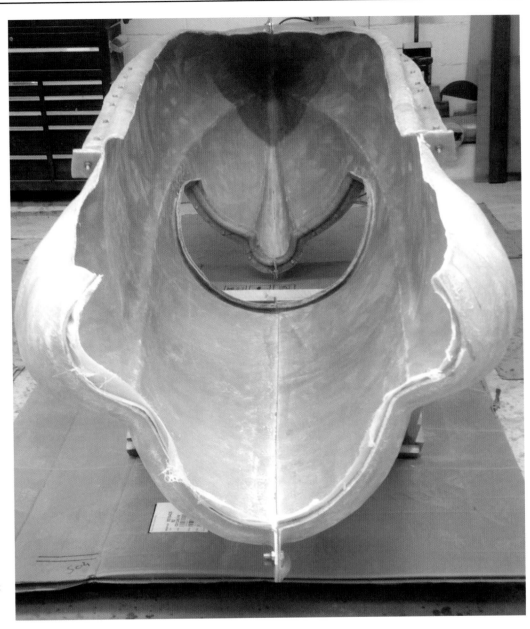

▼ The moment of truth. Avoid using anything metal to break the mould halves apart. A plastic wedge and rubber mallet are perfect. *(Master Mechanic)*

▼ The birth. This is the moment to be proud. You've just created a brand new body from scratch, after building the buck and mould. *(Master Mechanic)*

had been worthwhile taking time and care with the release-agent layer.

Once the mould popped, the process was not over. The body has a lot of curves and different angles that wanted to resist the mould being removed. I had to use brute force – the glassfibre is very strong and tough, but it's important to use only plastic or rubber tools to avoid damaging the gel-coat surface. I had to pull really hard (it might be necessary to get some friends to pull on each side), as I prepared to reveal the body to the world for the first time.

Just like a newborn calf, there was a sticky layer on the bodyshell that needed removing – the release agent. I could have licked it off, like a calf's mother does, but I decided to use a sponge and some warm, soapy water! While I was doing this, I inspected the body surface for defects, checking that the gel coat was regular, and that there weren't any thin spots. Any odd imperfections could be filled later.

There was still a raised gel-coat seam running down the centre of the shell, where the two mould halves were joined, and this had to be sanded off. Using a random-orbit sander and some fine-grit abrasive paper, I took the seam down to the level of the rest of the body. I used an air saw, then abrasive paper, to remove a few sharp edges.

The shell was starting to look pretty good, and I decided to reward myself with a beverage. I also decided to sit in the 'cockpit' and had a photograph taken while I held a washing-up bowl for a steering wheel!

Cooling louvres

When I'd finished claiming glory, I had some more glassfibre work to do. The engine compartment needed to allow fresh air in, and some of the engine heat out. For that, I needed to fit some louvres – look at any race car from this era and you'll see the same shark-gill-like slots in certain areas. I decided to produce these in glassfibre, and the technique was exactly the same as the rest of the body. I used twin louvres on each side of the car.

I made a cheap louvre mould using a household vent from a local hardware store as a buck. I applied the usual release agent to the vent, and laid up glassfibre to create the part that would go on the car. I made the flat section of each louvre bigger than it needed to be, so that I could cut it to shape as required.

Next, I marked the positions of the louvres on the bodyshell, and cut both sections so each louvre would slot in. I needed to create a hole in the body for each louvre to sit in, then I held the louvres in place with tape (a hot-glue gun would do just as well), and secured them in place using CSM and resin, as previously.

▲ It's back on the sander to remove the raised seam between the two halves. *(Master Mechanic)*

◄ The vents can be made using the same techniques as the main body. *(Master Mechanic)*

▼ Just like a shark's gills, these glassfibre louvres allow the beast to breathe. *(Master Mechanic)*

◀ **Taking a break to sit and stare at the fruit of my labours. I did a lot of that during this build.** *(Master Mechanic)*

Lower valance panel

I created a further glassfibre panel to act as a brace for the lower front of the bodyshell. I moulded this around some shapeable foam, making the panel oversize, and trimming it down when the material had fully cured. This would stay on the car as a lower valance panel, but also added a ton of strength when the bodyshell was being moved around.

Bonnet

The next job was to cut the bonnet shape from the bodyshell. I used a roll of narrow masking tape and got creative with some bonnet shapes on the bodyshell. I was looking to open up a hole big enough to access all the parts of the engine. I didn't need to worry too much about removing the engine through the bonnet, as the finished bodyshell would be removable – if the engine has to come out, the shell can come off first.

▶ **The finished body shell needs more work, including marking out for the bonnet to be cut away. Consider whether you might need to remove the engine in the future. My body was always removable, so the bonnet was really just for regular access.** *(Master Mechanic)*

I stuck to the age-old advice – measure twice, cut once. When I was happy with the shape defined by the masking tape (I used a similar shape to that on the original Alfa 158), I cut it out. I used a small pneumatic saw with a really narrow blade to get around the corners, though a jigsaw would work just as well. I took my time and wore full Personal Protective Equipment (PPE) for this, as the dust produced by hardened fibreglass is dangerous, and must be avoided at all costs.

▶ **Steady does it. I used a small pneumatic saw to cut the bonnet opening, but it needed a steady hand.** *(Master Mechanic)*

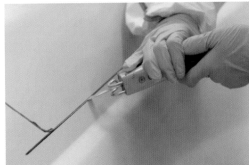

Fuel filler cap

The last detail to take care of on the bodyshell was the fuel filler cap. I anticipated where the fuel tank would go in the tail section, and cut a teardrop-shaped fuel filler cap. The shape needed sanding smooth at the edges. I needed a hinge for the flap, so I used a kitchen-cabinet hinge, which was perfect, because it extended as it opened to clear the hole.

Finishing touch

Finally, I gave my finished bodyshell a full inside coat of gel coat. I opted for black, so the interior and engine colours would pop. It's possible to be creative here and use any colour you like. The gel coat filled in most of the scruffy, rough glassfibre look, though I didn't obsess over it – it's a race car after all.

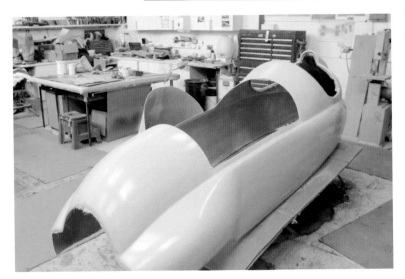

◀ **Pretty much the finished article. The bodyshell is lightweight, very tough and just about ready to be fitted to the car.** *(Master Mechanic)*

CHAPTER 16

Adding a space frame

With the foundations of the ladder chassis built, and the bodyshell in my possession, the next step was to make the two fit together perfectly.

The body was now sitting on timber furniture dollies to enable it to be wheeled around. It didn't weigh much, and could be moved around a smooth workshop floor easily. The front of the shell had the extra valance piece in place, that I made during the bodyshell construction, to strengthen and brace the nose.

The chassis was more difficult to move, and I placed it on four axle stands.

At this point, the body was one clean, smooth, continuous shape. In order to progress, I needed to mount the bodyshell over the ladder chassis, allowing room for the suspension components. This meant cutting holes in the four corners of the shell to accommodate the front and rear suspension. Before cutting, some planning was needed.

The starting point for my plan was to measure the height above the floor of the centrepoint of the front and rear wheels, with tyres fitted (for my build, the front and rear heights were subtly different due to using different-sized tyres front and rear). The centre of the front wheel was 13¼in from the floor, and the centre of the rear wheel was 14in from the floor.

Next, with the body resting on the floor, I worked out roughly where the front and rear wheels would be positioned laterally on the body (this lateral position does not need to be exact at this stage, the important

thing is to mark the heights of the wheel centres). I then measured up from the floor and made marks at the heights of the wheel centres front and rear (on both sides of the body), but I subtracted 4in from each measurement to allow for the intended 4in ground clearance (ride height). So, in my case I measured 9¼in up from the floor at the front, and 10in at the rear, and drew horizontal lines on the body.

It was now essential to set the chassis at the correct ride height. This would determine the final position for the body in relation to the chassis.

▲ **When finished I opted for a nice ivory cream-coloured powder coating.** *(Ant)*

▼ **Getting closer. With the cockpit location now fixed, I could work out where I would end up sitting.** *(Master Mechanic)*

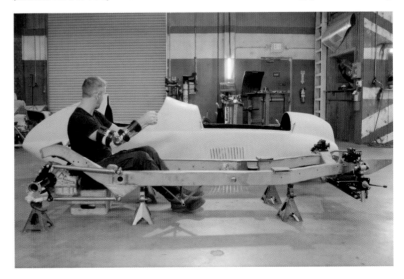

▲ **It's amazing how close the slightly modified MG chassis is to the bodyshell.** *(Master Mechanic)*

The chassis resting height above the floor is determined by the wheel/tyre size and the suspension travel.

I raised the rear axle on axle stands so that it rested with the centre of the stub axle at the correct height for the wheel centre (14in from the floor in my case). I then adjusted the axle stands supporting the rear of the chassis so that the lower trailing arms were horizontal – I used a spirit level to check this. This set the rear chassis height.

At the front end, I used cable-ties to raise the lower wishbones so the front spindles were 13¼in

▶ **I used a laser line for accuracy and marked my cut lines with a pencil.** *(Master Mechanic)*

▼ **Marking the wheel centres is the first stage in marking out the cuts for the suspension.** *(Master Mechanic)*

from the floor (front wheel-centre measurement). I then adjusted the axle stands supporting the front of the chassis, so the front lower wishbones sat horizontal (checked with a spirit level). This is roughly in the middle of the 'travel zone' for the double-wishbone set-up.

One last double check, and the chassis, rear wheels and front wheels were now positioned in their final resting place. This was the 'parked' position of my rolling chassis. Note that the chassis actually dips down, with the front lower than the rear.

Now it was time to lower the body over the chassis for the first time.

As I had a two-post ramp, I supported my body on just two of the arms and hovered it above the chassis. I then spent a great deal of time ensuring that both the body and chassis were aligned. Taking time over this stage meant that lowering the body using the ramp would give perfect alignment when the chassis and body were finally mated together. Even just a few millimetres off would have an impact on the way the final car looked on the road. The body HAD to be fitted accurately and straight!

The idea was to lower the body so the axle (wheel) centres aligned with the horizontal lines made previously on the bodyshell.

I hovered the body as close to the chassis as possible, so that it sat just above the correct position. It was important at this stage to ensure that the body was in the correct lateral position, so that it overhung the chassis correctly at the front and rear, with the tapered chassis fitting inside the taper of the body at front and rear.

Once I was happy with the body position, I used a laser-level to shine a vertical line on the body exactly in line with the position of the centre of the front stub axle (the wheel centre). I then marked a vertical pencil line on the body where the laser line passed through the horizontal line made earlier. The intersection of these two lines gave the exact position of the wheel centre on the body. I then repeated the steps again to mark the position of the rear-wheel centre. I carried out this measuring and marking at front and rear on both sides of the bodyshell.

These wheel-centre marks on the body provided a great starting point for my cuts to accommodate the suspension.

At this point, I eyeballed through the nose, and I could see that the chassis rails followed the tapered shape of the bodyshell. It's amazing that a 1950s two-seater MG TD chassis fits pretty closely to the shape of the 1930s 158 single-seater!

Next, was a slow process of measuring and marking the suspension points onto the bodyshell. I had already marked the horizontal and vertical lines for the centres of the wheels, so now I again used a laser line to mark all the critical vertical points around the suspension, and I made pencil marks on

▶ **At the rear I used the laser line to mark the position of the suspension components on the chassis below, then I used the spare boomerang from the axle build to create a cardboard template. With the 'cross' (intersection of horizontal and vertical lines marking the wheel centre) positioned exactly in the centre of the circle formed by the axle 'cut-out' in the centre of the boomerang template, I lined up the vertical laser line so that it was touching the front edges of both boomerang arms, ready to draw a vertical line along the laser line. It's incredibly accurate.** *(Master Mechanic)*

the bodyshell corresponding to these points. I erred on the side of caution to keep the holes as tight as possible, knowing that once the body was mounted I could open the holes up to accommodate the suspension travel and allow breathing space around each part. You can cut material off, but you can't stick it back on!

I used the laser level for any vertical or horizontal lines I needed to ensure accuracy, for instance making sure the boomerang templates were vertical on the bodyshell.

After checking and double checking, I placed masking tape along the pencil lines to act as a cutting guide. Knowing I was likely going to have to make the hole bigger, I started with small cuts first, using a reciprocating air saw. This was a big moment!

After cutting all four corners, with some help to flex the bodyshell open, I then lowered the body over the chassis, until the body hovered 4in off the floor (my chosen ride height).

Next, I rechecked that the front and rear slots allowed for sufficient suspension movement, by lifting the axle at the rear and the stub axles at the front (this is a trial-and-error process), and with the body in its final resting position (in position over the chassis at the correct ride height) the next move was to temporarily secure the shell in place.

◀ **Once again, measure twice cut once. While everything is ultimately repairable when working with glassfibre, you really don't want to be fixing mistakes.** *(Master Mechanic)*

◀ **I applied masking tape along the pencil lines, then cutting against the masking tape with the reciprocating air saw, I cut the clearance slots for the suspension.** *(Master Mechanic)*

◀ **At this point I discovered that I needed to extend the rear clearance slots in the body forwards to allow sufficient clearance for the trailing arms, so again I marked the bodyshell, placed masking tape as a cutting guide, and used the saw to extend the slots.** *(Master Mechanic)*

An absolutely splendid graphic which shows the space frame fitted inside the shell. *(Noah Schutz/Paul Cameron)*

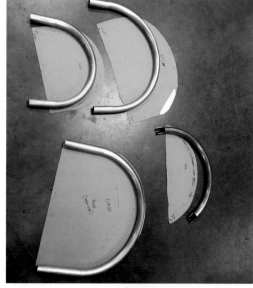

▶ The space frame starts with four curved hoops. Many metal fabricating companies have the kit to bend hoops like this. *(Ant)*

▶ Make sure the hoops have extra material on the ends, to make them easier to notch and fit. *(Master Mechanic)*

I now needed to add a space frame to the ladder chassis, so I secured the shell to the chassis frame using clamps, with thin wedges of steel between the chassis and the inside of the body, to allow me to move the ramp arms out of the way.

To create the space frame, I started with four main 'hoops' – a front hoop to support the nose, a front-bulkhead hoop, a midway hoop to support the cockpit bulkhead (and to mount the dash and the steering column) and a rear hoop to support the tail (and to mount the upper ends of the dampers and the fuel tank).

The shapes of these hoops are unique to the internal body shape, so I made cardboard templates for the hoops. I have a tube bender in my workshop that's perfect for bending tube, but not for these longer-curvature shapes. For that I needed a set of rollers. So, I took my card templates to a local tube-bending specialist and asked them to make the hoops for me. They are made from 1¾in ERW (Electric Resistance Welded) 14SWG tube – nice and strong to make my space frame safe, stiff and rigid.

The bodyshell now became a template – a jig – for the steelwork. The plan was to build the space frame within the inside walls of the shell, starting with the four hoops.

I started with the rear hoop first, notching half the base of the tube so it hung over the rear edge of the rear chassis crossmember. I then tack-welded it in place.

SPACE FRAME

In the days before monocoque construction became common, sports and racing cars produced in small volumes were often built using space frames. The space-frame concept was invented by Alexander Graham Bell (yes, the guy who invented the telephone) in 1898, initially to experiment with structures for maritime and aeronautical engineering. A space-frame structure relies on the inherent rigidity of a triangle, and consists of numerous lengths of tubing welded into a web-like structure. It really came to the automotive world via the aircraft industry, but has been used in everything from bomber aircraft to roof-trusses.

The 1954 Mercedes 300 SL Gullwing and subsequent Roadster were both built using space frames. The concept was used for many other small-volume vehicles, such as the Setra S8 bus – so light six men could lift the frame. This self-supporting design was developed by Otto Kässbohrer in 1951 and was widely used for many years. The design even gave the Setra brand its name, which comes from the German word 'selbsttragend', meaning self-supporting.

The space-frame idea was taken to extremes by the famous Maserati 'Birdcage', or Tipo 60 and 61, of 1959, which used 2- or 3-litre versions of Maserati's 4-cylinder twin-cam. Tubular space frames are time-consuming to build, which is why they only tend to have been used in low-volume racing cars or supercars, but they offer a great strength-to-rigidity ratio when built with the correct type of steel tubing.

▲ **It's very easy to see how the Maserati Tipo 61 acquired its 'Birdcage' sobriquet.** *(Maserati SpA)*

▶ **The bodyshell can be drilled easily, and the steel tabs on the hoop can be tapped with M6 threads to act as bodyshell mounting brackets.** *(Master Mechanic)*

Note: *Take suitable fire precautions when welding with the glassfibre bodyshell in place.*

I then placed two steel tabs (1in x 3in), made from 4mm mild steel bar, either side of the headrest, and I drilled and tapped them to take a 6mm bolt.

I placed another 4mm steel tab (approx 3in x 3in, but shaped) resting on the inside of the bodyshell, then gusseted to the inside of the ladder chassis, just above the rear suspension hole in the bodywork. I then drew a continuation line for the rear, angled line of the rear suspension hole, and again drilled and tapped 6mm holes, two either side of the line. The front two holes are for fixing the rear of the main body, the rear two for the front end of the tail section.

I then repeated the procedure to fit brackets on the other side of the chassis. With the bodywork secured to the brackets, I used an air saw to carefully cut up the pencil line on each side of the body to detach the tail from the main bodywork.

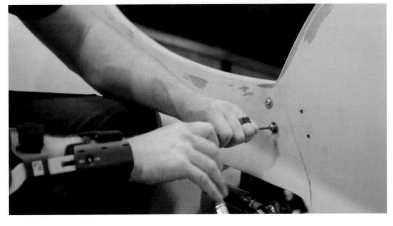

▶ **I used some nice-looking stainless, dome-headed Allen bolts and washers to secure the body. They add a nice touch to the exterior.** *(Master Mechanic)*

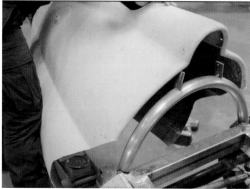

▲ **Lifting the bodyshell rear section from the space frame reveals the brackets welded to the hoop on either side of the headrest.** *(Master Mechanic)*

With the tail section of the bodywork removed, I could now complete the rear suspension.

I used coil-over-damper units at the rear end, and I elected to use GAZ dampers, with ½in rose-jointed eyelets at either end, and an open length of 14½in and closed length of 10½in.

Using a ½in bolt and some tapered aluminium spacers either side of the damper eyelet (to allow for the rose-joint movement on the damper), I bolted the bottom damper eyelet to the threaded bush on the rear axle, then bolted a loose threaded bush (the same as used in the trailing arms) to the top eyelet.

I removed the spring, and held the damper in the middle of its compressed/extended range. Holding it upright (aided by a spirit level), I joined the bush (previously bolted to the top damper eyelet) to the chassis with some 1¼in 14SWG steel tube to create an arm. I then triangulated that arm to the top of the space frame rear hoop with some 1in tube – all nice and compact!

I repeated the process to fit the coil-over-damper unit on the other side of the car, and that completed my rear suspension. Back to the space frame...

▲ **It seems sacrilege to chop the bodyshell up after all the hard work, but the air saw makes light work of it.** *(Master Mechanic)*

▶ **With the bodywork tail section removed, access to the entire rear end of the car is a lot easier.** *(Master Mechanic)*

▶ **Looking at where the rear coil-over damper units will mount. Space is not so limited back here.** *(Master Mechanic)*

▼ **The coil-over unit is fixed to the chassis with thick walled 1¼in tube welded to the rear hoop.** *(Master Mechanic)*

▼ **Here, I have welded the triangulation tubes between the damper mounting arms and the top of the rear space frame hoop.** *(Master Mechanic)*

The front hoop sits just inside the nose of the bodyshell. With the hoop already cut to shape using the cardboard template, it only took a second to tack it in place.

The front bulkhead hoop is also simple. I made it a snug fit to the bodyshell and ensured it was square to the chassis rails, again measuring from the front suspension for accuracy. I used a spirit level to make sure it was sitting upright. I had to remember that using a set-square against the horizontal chassis rail would not work, as the chassis sloped down towards the front. Again, I tacked it in place.

The midway (cockpit bulkhead) hoop was slightly trickier, as it needed to sit back at an angle to accommodate the dashboard (which leans backwards). I cut it to fit and tack-welded it in place.

With the four hoops in place, I moved on to securing the lowest part of the main bodyshell, creating supports to hold the lower 'sill' section. As previously, I used the body as a jig to position these supports correctly. I cut some 1in-diameter ERW 14SWG tube, 4in long, and some 4mm-thick, 1in flat bar. The 1in tube rested perfectly within the internal curvature of the lower bodyshell 'sill'.

With the main bodyshell in place, I measured from the bottom inner and outer edges of the lower chassis rail to the 1in tube (resting in place against the inside surface of the 'sill'). I then cut two sections of the 4mm-thick flat bar to match the measurements, which would form a bracket to secure the tube to the chassis.

I made four of these supports in total – two on each side of the chassis – to support the front, and middle of the main bodyshell (I will move on to the rear supports next). I welded the four supports in place, welding the bar sections to the chassis, and to the tube.

For the rear main bodyshell support brackets, again, I used 4in-long sections of the 1in-diameter

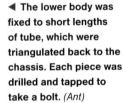

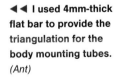

▲ The front hoop is tackled in the same way. The idea is to provide strong fixing points for the body wherever it has the ability to flex. *(Master Mechanic)*

◀ The lower body was fixed to short lengths of tube, which were triangulated back to the chassis. Each piece was drilled and tapped to take a bolt. *(Ant)*

◀◀ I used 4mm-thick flat bar to provide the triangulation for the body mounting tubes. *(Ant)*

◀ The finished mounts are strong and relatively light. The whole chassis assembly will be powder-coated to protect it. *(Ant)*

▲ **Getting there. It's starting to look like something you might see at a race track, minus a few things.** *(Ant)*

▶ **I cut a piece of 1¼in tube that was oversized and made a small bend in each end. I then held it inside the engine bay and marked where to cut it at front and rear.** *(Master Mechanic)*

▼ **I used a power saw to cut the tubes to length before welding them in place between the hoops.** *(Master Mechanic)*

tube, but this time I joined the left and right tubes together, with another section of 1in tube running horizontally between them, from left to right across the chassis. This would ensure that the body sat centrally. Using more 1in tube, I triangulated the supports to the chassis. I welded a length of tube on each side vertically, between the outer end of the support and the lower face of the chassis rail, and

another length running forwards to the lower trailing-arm mounting bracket.

Working at each support in turn, I drilled two holes through the body and into the support tube, using a 5mm drill, then tapped with an M6 tap. I fitted two stainless-steel bolts, ¾in long, to secure the body to each support.

With all the lower brackets in place, the bottom half of the body was firmly secured. The top of the body still had some play, and to resolve that I returned to the four space-frame hoops.

The next stage of creating the space frame was to join the hoops together with lengths of the same 1¾in round tube. The tubes in the engine bay require a small bend at the front and rear to allow them to sit tight to the body while at the same time reaching the hoops. Making the required bends in the tube was easy using a tube bender, but the same effect could be achieved by cutting and welding the tube at an angle, though that is not as aesthetically pleasing!

I rested the tube so that it naturally fitted within the bodyshell swage line below the bonnet opening.

The body again acted as a jig to hold the tube in the correct place for tack-welding. I tacked the tube to the front hoop and front-bulkhead hoop, and then repeated the procedure to fit a tube on the remaining side of the chassis.

At the rear, I joined the angled dash hoop to the outer edge of the MG TD chassis rails, creating 'arm rests' for the driver (and some side cockpit protection), also bracing the space frame.

I decided to fit a 'Brooklands- style' aeroscreen (available with a curved mount for an MG TD) – very similar to the screen used on the original Alfa 158. I placed the screen on the bodyshell, in front of the cockpit, and when I was happy with its position, I drilled through the aeroscreen mounting holes into the shell, creating four holes. I then cut four 1in x 3in tabs of 4mm steel to cover the holes from the underside, and to reach to the edge of the midway (dashboard) hoop. Next, I drilled and tapped a 6mm hole in each tab, screwed the screen to them to hold them in place, and welded them to the midway hoop. These tabs would not only mount the screen later, they would also firm up the bulkhead while hiding the screen-mounting bolts.

The remaining bulkhead metalwork would be finished later, when the bodyshell was removed to expose the chassis.

Returning to the bodyshell tail section, I carefully opened up the axle hole to accommodate the coil spring/damper units that had now been fitted, and I then refitted the tail section to the chassis.

At this stage, I had to build a frame to support the tail section. To hold the base of the tail section, I first welded a 'V' on the workbench from 1¼in 14SWG round tube. I then cut a triangle from 4mm steel plate to fit inside the base of the tail. Next, I rested the 'V' inside the tail, with the open (front) end of the 'V' against the top of the rear edge of the rear chassis crossmember, and the front of the 'V' resting on top of the steel-plate triangle (which was still resting in the base of the tail). I leaned carefully into the tail and tack-welded the 'V' and triangle in place, taking great care not to damage the bodyshell, and taking plenty of fire precautions! I then welded a short piece of steel tube to provide triangulation between the 'V' and the front of the triangle.

Next, I made two supports for the lower edge of the tail section. For these, I used two sections of the same 1in-diameter ERW 14SWG tube that I used for the main bodyshell supports to the 'sills'. This time, instead of using flat steel bar to support the tubes, I used the 1¼in 14SWG round tube that I used to build the tail-section frame. Working on each side of the chassis in turn, I rested the 1in tube in the internal curvature of the lower part of the tail section, then cut a section of the 1¼in tube to join the 1in tube to the previously fitted 'V' frame. As with the main bodyshell brackets, I then drilled two

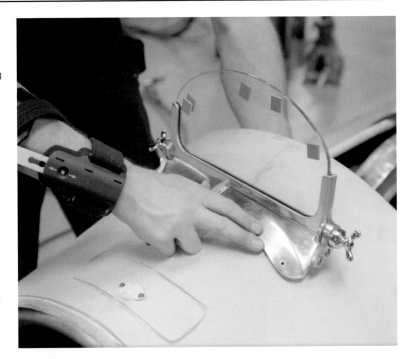

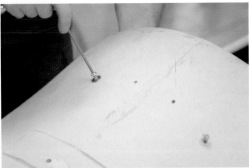

▲ A very cool 'Brooklands-style' aero screen. This adds such a great period feel to the car. *(Master Mechanic)*

◀ The aeroscreen is fitted using the same technique as the body mounts. Drill, tap, screw. *(Master Mechanic)*

◀ The body rear end fully fitted. Note, the extra cut-out space for the coil-over damper units. Trimming the body is a bit like hairdressing – you can take more off, but you can't add it back. *(Ant)*

▶ **The tail structure. I used the lightest material I felt I could get away with. It needed to support the fuel tank, the rear body section and itself.** *(Ant)*

▶▶ **Working in this tight space was something of a challenge. As soon as things were tacked in place, I could remove the bodyshell and weld fully.** *(Chris Hill)*

▶ **My damaged arm was still trouble. It didn't hurt, but mobility and holding things in place was tricky.** *(Master Mechanic)*

▶▶ **Magnetic squares like this are great for holding steel tubes when preparing to weld them.** *(Master Mechanic)*

holes through the body and into each support tube, using a 5mm drill, and tapped with an M6 tap. I fitted two stainless-steel bolts, ¾in long, to secure the tail section to each support.

With the newly built tail support-frame components tack-welded in place, I removed the tail to improve access.

Now that I had easy access to the rear of the chassis, the next step was to add some triangulation using ¾in round tube.

I formed a lower dash piece on the tube bender, using a piece of the 1¾in tube, to fit across the midway (dashboard) hoop. I then drilled a 1in hole at the top centre of the tube, and welded a ¾in nut to hold a ¾in rose joint (used later to support the steering column). Working on each side in turn, I welded the tube in place at the junction point of the hoop and the side tube.

◀ **Lifting the bodyshell on and off is definitely a two person job. You can see how thin and flexible the whole thing is. Lift, Chris, lift.** *(Master Mechanic)*

It was now time to enlist some help, remove all the bodyshell securing bolts and slide off the main bodyshell to reveal the chassis work so far.

I added more 1¾in round tube between the front-bulkhead loop and the midway (dashboard) hoop, and then some ½in tube underneath to triangulate this section further.

Where the bulkhead hoops were overlapping the outer edge of the ladder chassis, I cut shaped gussets from the same tube and welded them in place for reinforcement, and to provide the finishing touch!

Next, I added triangulated support to the front of the front hoop, using ¾in tube, then using ½in tube I triangulated between the engine-mounting plates on the ladder chassis and the front and front-bulkhead hoops.

Later on, I added Weber sidedraft carburettors to the engine, and I found that I had to make some modifications to the right-hand side of the engine bay to accommodate them. I had to remove the tube connecting the front-bulkhead hoop to the midway hoop, as this would foul the carburettors. I decided to replace this tube with two tubes (with bends), fitted between the hoops and the ladder chassis to clear the carburettors.

That was the space frame completed, so I then set about fully welding up the entire chassis.

Before committing to powder-coating the chassis, I continued my build, knowing that I would need to add brackets later (to support the wiring harness, brake lines etc). This was, after all, an ever-evolving build.

Now for the cockpit.

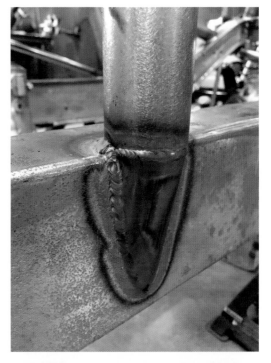

◄ The bottom of a bulkhead hoop. The gusset was made separately from the same tube and welded in place to provide reinforcement. *(Chris Hill)*

◄ After adding Weber sidedraft carburettors, I had to make modificatios to the space frame to accommodate them. Note the two curved tubes either side of the carbs. *(Chris Hill)*

◄ It's worth enjoying these moments, even if it means using a bowl as a pretend steering wheel! *(Chris Hill)*

CHAPTER 17

Creating the cockpit

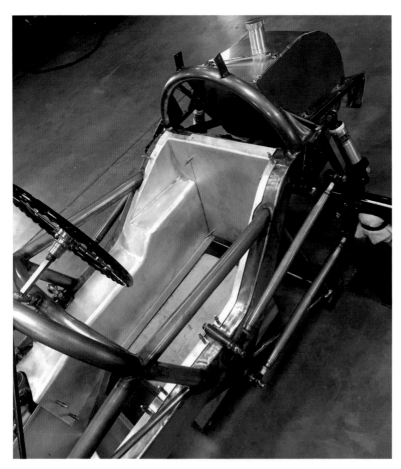

I wanted my car to have a vintage race feel, and nothing says 'race car' more than a beautifully crafted bare aluminium cockpit. My plan was to create a simple tub that slotted into the chassis, creating a space for me to sit. I decided not to install a seat, and instead I would trim the space to provide the appearance of a seat, utilising every available millimetre of space in the cockpit.

There are no rules when creating a cockpit like this. My aim was to give myself as much room as the car would allow. The final seating support and position would be finely tuned using the upholstery.

To create the cockpit tub, I needed to work around various essential components. At this stage, I didn't yet have my engine back from the machine shop, but as I had my exchange gearbox ready, I could install that and still make progress (one of the benefits of already having decided on the gearbox-mounting positions). With the gearbox, rear axle and propshaft in place, I could work around them to build my cockpit tub.

To help visualise my seating position, I had been using wooden apple boxes to position myself as I progressed with the build – a 9in-high box for my bum and a 4in-high box for my feet. Lowering the position of my feet (and the pedals) bought some much-needed extra legroom.

I made the cockpit tub from flat sections of 2.5mm aluminium. I used a marker pen and a spirit level for the vast majority of the marking out. I wanted the

▶ **The driver was often an afterthought in early racing car design, squashed into the remaining space. Offsetting the engine/ gearbox allowed me to have a larger cockpit, as ultimate performance was not the main priority.** *(Paul Cameron)*

bottom seat panel to be parallel to the road, and also the tub section directly below the pedals. The spirit level doubled up as a great straight edge.

A sheet-metal folder was invaluable here – this puts folds into the 2.5mm aluminium sheet with ease. There is the option of making every corner a cut/weld, but a sheet-metal folder avoids all that. I'm lucky enough to have one in my workshop.

Some sheets of stiff card and scissors were essential for making templates. I started with the left-hand sidewall, creating flat panels from the aluminium sheet.

First, I cut a piece of aluminium and folded a 1in, 90° return to the lower edge (this was for the floor to eventually rest on). Hovering it 9in off the floor (on top of the wooden apple box) I marked the intersection point where the chassis narrowed, and cut a small slice into the 1in lower return, allowing me to place a small crease to follow the chassis line. I drilled a hole in the aluminium sheet to line up with the upper trailing-arm mount, so I could use the trailing-arm securing nut to hold the sheet in place. Notice I also left a small section at the front, where the chassis was flat, and folded the sheet outwards to snugly fit on top of the chassis rail. With the sheet held in place, I marked the profile of the chassis rail on the outside of the sheet with a pen. I also drew around the upper trailing-arm mount.

I then removed the panel and cut it along the pen lines to fit the chassis.

I wanted the flat panel to follow the chassis shape, so I cut a piece of aluminium sheet ¾in wide, and shaped it to fit along the curved top surface of the chassis. Welding that strip in place created a 3D shape, and the tub started to develop.

I continued the left side panel down towards the footwell, using the same technique of folding the bottom 1in edge over inside the chassis to support the floor. I cut the panel slightly long, marked the upper chassis shape, and again welded the top edge to the chassis.

Any seams created needed welding, but I just tacked everything together first, only completing the seams when the full tub was finished.

As I worked along, the growing panel tended to distort as it got heavier, so I used some clamps to hold it to the chassis. As I progressed, I also used some reusable fixings called Clecos. They act like removable rivets. I drilled a hole through the new panel and the chassis and inserted the Cleco with the appropriate tool, which then held the panel tight. When I was happy with the whole tub, once I removed the Clecos, the rivet holes were already there!

I left the footwell unfinished at this stage, until the pedals were fitted. The last thing I wanted was to find that the pedal movement was restricted by the aluminium panels I had welded in place. I continued with everything else.

▲ **The first piece of aluminium sheet clamped to the chassis for a trial fit. Note the return on the lower edge.** *(Ant)*

◄ **The flat top plate was cut and welded on as a separate piece.** *(Ant)*

▼ **I'm wearing a glove here, partly because I'm welding the pieces, but also because they can have sharp edges after being cut.** *(Master Mechanic)*

▶ **Slowly adding pieces to the cockpit as it takes shape.** *(Ant)*

▶▶ **Building first in component form, then welding to give a smooth shape.** *(Ant)*

▶ **Making card templates first helps to ensure a good fit before committing to aluminium.** *(Ant)*

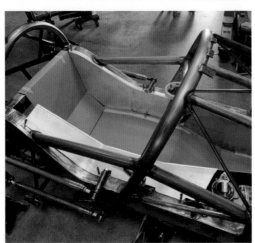

The right-hand side was a little trickier, as it involved forming the transmission tunnel. Instead of using a sheet-metal folder for the long corner, I had a 2in rolled edge produced for me at a local workshop. This gave a nice, transmission-tunnel look. I had to be careful to leave enough space for the gearbox on the right, making sure the pedal box would fit in the space I created. I sat on my wooden boxes and tested out the space numerous times before I committed to cutting metal.

I formed the right-hand side in the same way as the left. Working from the top down, I held the formed pieces in place as I worked along.

Building in component form then riveting together would make a strong and solid tub. The trick is to create a tub that slots in to the chassis and maximises the space, clearing the important components, yet resting solidly against the chassis.

Fitting the floor added a great deal of strength, in the same way fitting a hardboard back to a flimsy set of shelves adds a load of rigidity.

When I could finally sit on the floor, I stopped for another beverage and took a photo. Without the seat, I felt like I was sitting a little bit low, so I sat on a cushion to gauge how much extra height I would need for the seat.

I put in a few rivets at this point, to hold the tub in place, then it was time to drop the body back on.

Next, I needed to make a rear firewall panel. I cut this close to the fibreglass body, as the upholstery would attach to the body to cover the gap. I cut two slots in the rear firewall panel for the seat belt straps.

The whole tub could now be finished with the dual-action (DA) sander. I tidied up the welds – nobody's perfect – and there was a certain amount of 'making good'.

The tub, however, just provided the foundations for the cockpit – the detail was yet to come.

▶ **This is the perfect time to consider mounting points for a harness. The mountings must be strong and secure, and must meet the relevant regulations. Depending on the design, it may also be necessary to cut holes in the aluminium sheets big enough for the webbing to pass through.** *(Ant)*

CHAPTER 18

Cockpit controls

When building a special, it's essential that the car is bespoke and tailored to your exact needs. In a driving sense, the ergonomics are critical to the enjoyment of the car, especially a race car.

I love the simplicity of classic cars – no excessive display units, sat nav, stereo, or rows of switchgear that make modern cars look like spaceships. Nowadays, manufacturers spend millions of pounds developing the ergonomics of any new car. Every aspect – from the size of a cup holder to the feel of the trim, and even the way the car smells – is analysed in minute scientific detail.

Such 'progress' seems to have ironed out the rawness of earlier cars. For me, cars like the Ford Escort Mk1 Mexico or the original Mini Cooper were the essence of an engaging car to drive, allowing the driver and car to become one. There is a genius simplicity to those classic cars that's been lost today. Classic simplicity was the feeling I wanted to achieve for my car, and to do that I had to focus on three critical controls – the steering, pedals and the gearchange.

Steering

Starting at the front, I invested in a new steering rack and rod ends. The original rack was tatty and made a rather alarming clunking noise. A new rack was easy to get hold of and available as an off-the-shelf item, as were the rod ends.

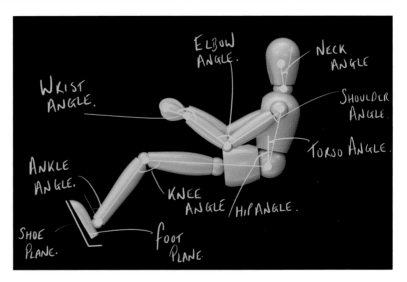

It's worth mentioning that as my donor MG TD was left-hand drive, a left-hand drive steering rack suited my offset engine layout. If I had started with a right-hand drive donor, then I would have considered using the left-hand drive steering-rack option.

I built my steering column using simple stainless-steel universal joints (UJs) to route it around the engine. To start, I cut off the top 1½in splined and threaded section from the pinion shaft on my steering rack, using an angle grinder.

◀◀ **An off-the-shelf steering rack for my Alfa Spider donor was cheap and easy to find. Left- and right-hand-drive options are available.** *(Chris Hill)*

◀ **I cut the splined and threaded end off the steering rack pinion shaft, as I didn't need it.** *(Chris Hill)*

► The universal joint at
the rack end had one
round, splined hole,
and one Double-D hole.
(Chris Hill)

►► The Double-D
('DD') hole fits the
matching shaft profile
and prevents the shaft
from rotating inside the
joint. *(Chris Hill)*

► + ►► I drilled a
shallow hole into the
shaft to locate the grub
screw, and then welded.
(Chris Hill)

My first UJ had a ¾in round spline at one end, and a ¾in 'DD' (Double-D – to fit the steering rack pinion shaft which had two flats) at the other. Two more 'DD' UJs would be required later. I slid the round end of the UJ over the rack pinion shaft, tapping it lightly so it was a nice snug fit. For extra security, I marked through the grub-screw hole with a pen, removed the UJ and then drilled into the steering rack pinion shaft to a depth of around

4mm, creating a locating hole. I refitted the UJ and inserted the grub screw. As a belt-and-braces exercise, I then TIG-welded the UJ in place on the pinion shaft.

Once the components had cooled, I fitted the steering rack to the car with the four securing nuts and bolts, and connected the rod ends to the uprights. That was the bottom end of the steering system done, next was the top end.

► The rack, and the
modified pinion shaft,
with universal joint
in place, fitted to the
chassis. *(Ant)*

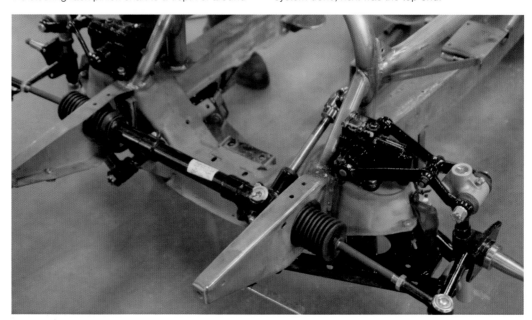

Like most cars from the period, the original 158 had a wood-rim steering wheel. Instead, I opted for an MG (Bluemel) steering wheel as a subtle nod to my original MG donor car.

It's amazing to see that racing cars of the period always had huge steering wheels. This was simply to give the drivers a mechanical advantage, to allow them to manhandle the cars around a track (with no power-steering systems to assist).

As the cockpit was rather tight, I added a quick-release steering-wheel boss to aid me getting in and out.

Starting with an MG TC aluminium steering-wheel boss, I first cut off the top 20mm (including the flange) using a saw. I then faced it off in the lathe to give a perfectly flat and true surface.

Next, I stripped down an off-the-shelf aluminium quick-release steering-wheel-boss kit and TIG-welded

◄◄ With no power steering on old racing cars, the steering wheels are often huge to increase the mechanical advantage for the driver. *(Ant)*

◄ My chosen steering wheel – from an MG. *(Ant)*

◄◄ I don't destroy everything I touch, I promise, but I did need to 'modify' the MG steering boss. *(Ant)*

◄ Having a lathe is very handy. It doesn't get used a lot, but at times it's a life saver. *(Ant)*

◀ **I used a standard steering wheel quick-release kit. With a bit of 'adjustment' the quick-release section worked perfectly for me.** *(Demon Tweeks)*

▲ **The flanged hub fitted my modified MG steering-wheel boss well.** *(Ant)*

▶ **Sometimes the answer lies within the parts you already have, but that might not be instantly obvious. You have to play around until the solution presents itself.** *(Ant)*

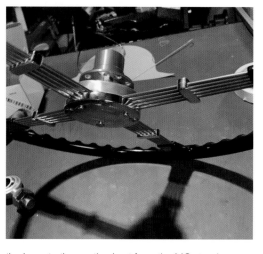

▶▶ **The splined end from the kit needed to fit a regular ¾in DD shaft. The lathe and the TIG welder earned their keep here.** *(Ant)*

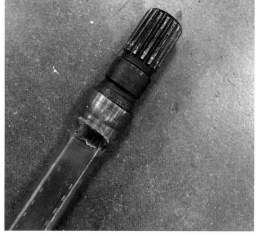

▼ **The steering wheel height was ultimately set by a rose joint on the front bulkhead hoop.** *(Master Mechanic)*

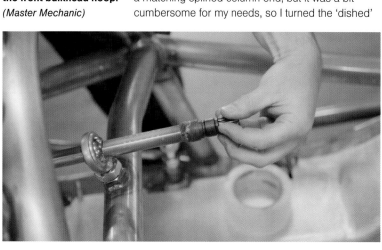

the boss to the section I cut from the MG steering-wheel boss. I then polished the assembly to finish it off. Finally, I bolted the steering wheel to the boss.

The quick-release assembly was supplied with a matching splined column end, but it was a bit cumbersome for my needs, so I turned the 'dished'

section off on the lathe, and drilled a 10mm hole into the centre of the column end (again on the lathe).

Next, I turned down the end of a length of ¾in 'DD' steering shaft to 10mm diameter, allowing me to slide the splined section from the kit onto the shaft, before TIG-welding them together.

Once the components had cooled, I fitted the steering wheel and boss onto the splined shaft end, before sliding the shaft end into the ¾in rose joint that I had fitted earlier on the lower dash chassis tube which I had welded in place on the midway (dashboard) space-frame hoop.

Next, I climbed into the car, and sitting on the aluminium floor panel (I had yet to fit the seat), with the steering shaft in place in the supporting rose joint, I held the steering wheel in a comfortable position to select my desired steering shaft position. There was a huge amount of play to enable adjustment of the rake angle up and down. I eventually found the perfect position to suit me, and then secured the shaft using a second rose joint mounted on the front bulkhead

hoop. Using another ¾in rose joint, I welded the ¾in locknut to the 1¼in tube, then welded that to the front bulkhead hoop, and triangulated with some ½in tube for strength. The length of the tube that the rose joint was mounted on was determined by the position I wanted for the steering shaft. Once finished, I passed the upper steering shaft through both rose joints.

Allowing sufficient length for the upper steering shaft to enter the engine bay (and pass through a bulkhead I was yet to make) I cut it down, added a UJ to the bottom end, and drilled holes for the UJ locking bolts, knowing I needed my steering shaft to avoid the engine. That was the upper section of the steering shaft fitted.

With the upper shaft done, and the rack position

already fixed, I needed to join the two. Obviously I couldn't join them in a straight line, as the engine would be in the way, so I needed to turn through an angle.

With the exhaust planned ahead, I welded a third ¾in rose joint onto the chassis side rail, 20in back from the front-axle line. I cut my additional two DD steering shafts to the correct lengths, and fitted them between the upper shaft and the steering rack via two UJs. As with the UJ used on the steering-rack pinion shaft, I marked through the grub-screw holes and drilled the shafts to a depth of 4mm, then secured the UJs with the grub screws. This time I didn't weld the UJs to the shafts, just to allow for any future adjustments and in case I needed to disconnect the shafts.

▲ **I welded the nut to the tube, welded the tube to the hoop, and triangulated.** *(Ant)*

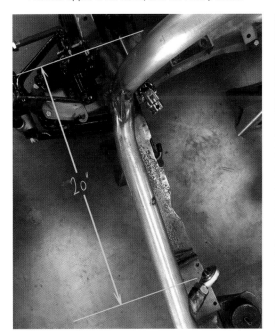

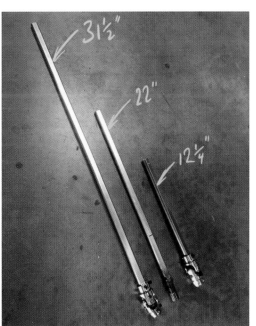

◀◀ **I welded a third ¾in rose joint onto the chassis side rail, 20in back from the front-axle line.**
(Ant/Paul Cameron)

◀ **The lengths of my three DD steering shafts were as shown.**
(Ant/Paul Cameron)

▶ The steering shaft had to sneak past the engine. I believe cars become much easier to understand when you see things simplified like this. *(Ant)*

▶▶ The routing of the steering shaft through the engine compartment on the finished car. Note the shape of the exhaust manifold, which had to avoid the steering. *(Ant)*

▶ The gear-lever linkage used many similar parts to the steering, although I had to get a bit more inventive with the way they were used. *(Ant)*

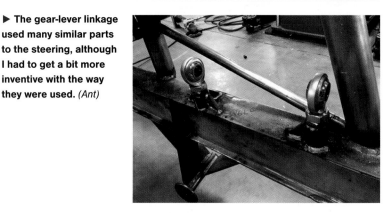

▶ The dimensions for the gear-lever linkage DD shaft with a tab and threaded bush welded on the end to act as a push/pull lever. The rose joints allow it to rotate as well. *(Ant)*

Later on, with the engine and exhaust in place, it all made sense. It was all a tight fit, but the 'quick' steering rack and larger steering wheel make light work of the action. Finally, with everything in place, I double-checked that all the components moved freely and allowed sufficient clearance for the movement of the engine and surrounding components.

Gear-lever position

The second phase of figuring out my cockpit ergonomics was to set up the gear-lever position. I could have left the gear lever where it was on the donor gearbox – it was a stretch, but I could reach it. However, I was not willing to compromise – I wanted my car to fit me like a glove, so the gear lever had to be relocated, and I opted to use a bespoke gear linkage.

Building a gear linkage isn't as daunting as it sounds. With a classic 'H-pattern' gear-lever gate I only needed to translate the movement forward and back, and left and right. I've seen countless gear linkages on cars where the positive, short-shift gear feel we all love is lost due to wobbly linkages and over-complicated designs.

First, dealing with the forward and back movement, I made use of the same ¾in DD material I used for the steering shafts, and I decided to make rigid mountings using more rose joints. I made two mountings for the rose joints, by welding a ¾in nut to a 1¼in-long piece of 1¼in-diameter tube. I then screwed the rose joints into the mounting tubes, and welded the tubes to the chassis, one 4in from the gear lever and the other 13in. I added 4mm triangular-shaped gussets to the front and rear of each tube to ensure that they were completely rigid.

Using a length of DD steering shaft, I welded on

◀ **The existing gear lever emerging from the gearbox needed a ½in bolt welded on.** *(Ant)*

◀◀ **Two rose joints joined together became a link which fitted to the gearbox lever.** *(Ant)*

a tab and a threaded bush with a ½in internal bore (the same as I used for the rear-suspension trailing arms). The bush would house a pivot bolt. Inserting the DD shaft through both rose joints would ensure that it slid back and forth precisely.

I took two ½in rose joints – one male and one female – and joined them together using a locking screwed onto the threads of the male joint. This creates a link.

I then TIG-welded a ½in bolt to the top of the existing Alfa Romeo gear lever (emerging from the gearbox), leaving enough thread on top to accommodate a nut.

Next, I slid a sleeve of ¾in steel tube as a spacer over the new gear-lever bolt, and after fitting two aluminium tapered spacers (the same as those used on the rear coil-over dampers), I used a nut to secure the female rose joint to the top of the gear lever. Using a ½in bolt, I fixed the male rose joint to the threaded bush on the DD sliding bar.

I then moved the DD bar, so that the link (the double-rose-joint assembly) was positioned at 90° to the gearbox front-rear centreline (bearing in mind the gearbox is set at an angle). It was essential that the rose joints were positioned at 90° to the gearbox front-rear centreline and directly in line with the left–right neutral plane of the gear-lever gate. I then welded the moving balljoint of the male rose joint to its own casing in that exact position.

The combination of the raised bush on the DD selector shaft, one fixed (angled) and one moving rose joint, and the ability to rotate the DD selector shaft provided the full forward and back, and left and right, H-pattern movement at the gearbox. I tried it for the first time, and it was quite precise and satisfying.

The great thing was that the whole assembly was adjustable at every angle, so I could fettle it until it was spot on and slid through the gears perfectly. I found that lowering the forward chassis rose-joint mounting a few turns, so the DD selector shaft sloped down slightly, really helped.

Using an off-the-shelf DD collar, I slid it onto the cockpit end of the new gear selector shaft, and again I sat on the floor of the car and chose my new gear-lever position. The plan was to find a comfortable resting position for the lever, and still allow free movement. I needed to be sure that I could move the lever to get into fifth gear without banging the chassis,

◀ **The linkage worked first time, just. The beauty of the adjustability that this set-up provided was that I was able to sit in the seat and really dial it in to suit me.**
(Ant/Paul Cameron)

▶ The last part of the gear-lever mechanism was the lever itself. Again, I built in some adjustment so it could be fine tuned. *(Ant)*

▶▶ I opted for a simple grey gear knob, but you could get creative. Not a pool ball though. I think that's been done enough! *(Ant)*

▶ The finished gear-selection mechanism doesn't look much like a regular road car, but it is a joy to operate. *(Ant)*

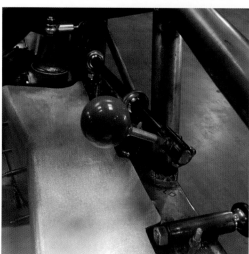

and equally I had to be able to reach reverse. Once I was certain of the position I wanted, I had to decide on the height and location of the gear knob itself (the height of the gear stick and its fore-aft position on the DD selector shaft). I held a 5in-long, 10mm bolt against the collar to gauge the height, and once I was happy, I tack-welded it to the collar and locked the collar in position with the supplied grub screws. After checking again, I fully TIG-welded the bolt to the collar, then drilled holes in the DD selector shaft for the grub screws and refitted the collar assembly.

I then purchased an aftermarket classic, round gear knob, with a 10mm thread, and screwed it onto the top of the assembly.

I tried the gear-selection action again, and it was amazing how slick it was with the leverage and comfort of a proper gear lever.

Pedals

The third, and final, part of the ergonomics trilogy was the pedals. Instead of raiding the donor Alfa Romeo Spider, I opted for an off-the-shelf universal pedal box. These are available in either floor-mounted or 'pendulum'-mounted configurations. I went for the pendulum-mounted option so that I could mount the assembly on the higher chassis structure, and not the aluminium cockpit tub.

I purchased a full pedal kit, which came with master cylinders, brake-bias bar, bias adjuster for the dash, brake fluid reservoirs, reservoir mount, and braided fluid hoses. It's a comprehensive kit and great for a special build.

The front–rear brake bias control is a really sneaky bit of kit. An adjustable bar in the pedal box alters the pedal pressure supplied to the pistons in the front and rear brake-circuit master cylinders (the brake hydraulic system is split front–rear). This is used in conjunction with the proportioning valve to fine-tune the front-to-rear braking bias. The control knob is fitted to the dash.

The first stage in fitting the pedal assembly was

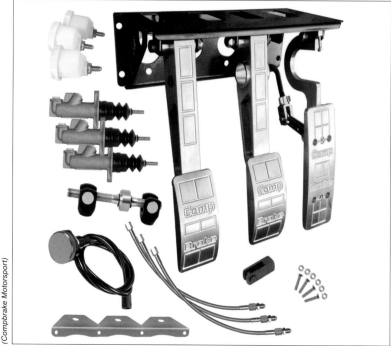

(Compbrake Motorsport)

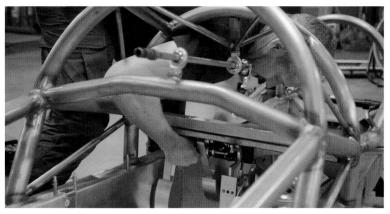

▲ Space became quite tight, but that's to be expected. Working out how it's all going to fit is part of the puzzle. *(Master Mechanic)*

▲ The dimensions of the steel frame I used for mounting the pedals, built from ¾in square tube. *(Ant/Paul Cameron)*

▶ The bulkhead and pedal-box cover panels not only seal the cockpit from the engine bay, but also provide a platform on which other components can be mounted. *(Chris Hill)*

to make a simple mounting frame from ¾in square 16SWG tube to straddle the MG TD chassis rails.

Spanning from the left chassis rail to the right, the mounting frame is supported on a pair of front legs and welded to the bulkhead hoop at the rear. I added two strips of 5mm thick, 1in steel bar between the upper rails of the frame, and from these I hung the pedal box within the open-topped aluminium footwell, using four bolts (the pedal box is supplied with pre-drilled mounting holes). I made sure the face of the pedal box was flush with the face of the square frame. It really was that simple.

At this point, I quickly climbed back into the cockpit and sat on the floor to double check that I was happy with the final driving position.

With the pedal box secured, it was time to carry out final fitting of the remaining cockpit-tub aluminium panels to the footwell, so that they finished flush with the end of the pedal box.

After making a cardboard template (with the body fitted), I then folded an aluminium cover panel to go over the pedal box front and top. I also made a new aluminium bulkhead panel.

▼ Three pedals and space for my feet is all I desire! *(Ant)*

CHAPTER 19

Brakes, clutch and hubs

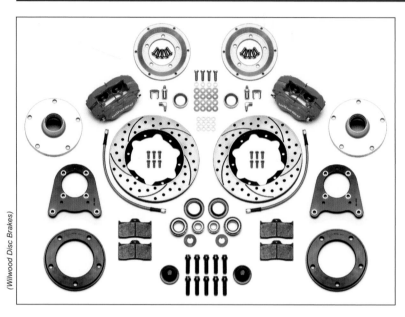

(Wilwood Disc Brakes)

Great brakes are an essential part of building a car like this. Pretty much every racing driver learns the phrase 'slow in, fast out' at an early age. This is the art of getting on the brakes late and hard (compared to driving on the road), reducing speed, turning into the corner, while balancing the car on the throttle, hitting the apex, and exiting the corner as quickly as possible.

A racing driver's sole job is to get around the track as fast as possible, and the corners only get in the way. The more efficiently the driver can slow, turn, and accelerate the better, conserving

▼ **This lovely panning shot of Farina at Silverstone in 1950 clearly shows drilled brake drums.**
(NMM Beaulieu)

momentum all the way. Of course, skilled drivers go on to develop a 'fast in, faster out' technique, as they perfect their craft.

Cars of the Alfa 158's period were fitted with drum brakes at all four wheels. The original Alfa 158 had hydraulically actuated drums of 14.8in diameter at the front and 13.8in at the rear. Perhaps if the car had been redeveloped for the 1952 racing season, which Alfa did consider doing, it may have been fitted with disc brakes. My 1952 MG TD donor also had drum brakes all round.

With non-assisted brakes, drums can sometimes actually perform better than discs. Drums have a 'self applying' characteristic, or are 'self energising'. Because of the pivot point and the rotation of the drum, the shoe is dragged onto the friction surface, making them grab more. The major disadvantage of drum brakes is that they do not dissipate heat as efficiently as disc brakes. This can be an issue for track use, where drum brakes can be prone to fade due to overheating.

Overall, disc brakes are superior, as they are able to dump heat far more efficiently, and this is particularly true with power-assisted brakes. I decided that my car would have four-wheel disc brakes.

With the pedal box fitted, next I fitted the master cylinders (including the clutch). The pedal box I used was designed to be used with separate master cylinders for the front and the rear brake hydraulic circuits, and also featured a built-in adjustable bias bar to allow adjustment of the brake bias between front and rear.

I located my three fluid reservoirs (again, a separate one for the clutch) on the bulkhead panel, using the bracket supplied in the kit. As the master cylinders are gravity fed, it was essential that the reservoirs were located higher than the master cylinders. The kit came with flexible braided hoses to connect the reservoirs to the master cylinders.

It is essential for the front brakes to provide a higher braking force than the rear. The role of the brakes is to slow the car as much as possible without locking the wheels – if a locked wheel is sliding across the road surface, it is not providing any grip. As a car brakes in a straight line, the front of the car dips, and the rear lifts, due to weight transfer – you can feel it. If the braking force applied to the front and rear of the car was equal, the rear wheels would lock as the load on them reduces. The trick is to dial in the correct bias between front and rear, so that the braking effort can be maximised without any of the wheels locking. In summary the rear-wheel brakes **have** to be less powerful than the front. I fitted new brakes all round, but I only made the front brakes power-assisted.

Hydraulic-fluid pipes

There is no set layout for hydraulic-fluid pipes, but there are some important rules. Metal pipes need to be securely fixed to the vehicle (I will use rubber lined 'P'-clips) and contact should be avoided with anything that moves or vibrates (such as the engine or gearbox). I obsess over clean, straight hydraulic-fluid lines – even the best cars can look average with wobbly or wonky lines, so I always like to make them neat.

I used ³⁄₁₆in copper and nickel pipe. Hydraulic-fluid pipes made from 90 per cent copper with 10 per cent nickel is becoming increasingly popular due to their superior corrosion resistance over pure copper.

Clutch hydraulic line

Starting at the clutch master cylinder, I ran nice, neat curves and lines from the master cylinder to the inside of the chassis rail above the slave cylinder. I made sure I followed the aluminium bulkhead panel closely.

Using a hydraulic-pipe flaring tool, I fitted the correct flared ends and fittings to join one end of the pipe to the master cylinder, and the other end to a new 7in (braided) flexible hose. I welded a small steel tab to the chassis to fix the hose position, then I joined the pipe to the hose, and the hose to the slave cylinder. To finish, I used 5mm rivets to secure aluminium P-clips, with rubber grommets, to the chassis to secure the pipe.

Rear brakes

Starting at the rear-hydraulic-circuit master cylinder, I ran a fixed pipe down the outside of the left-hand chassis rail. The bodyshell formed a handy cavity in this area, creating a nice space for brake lines.

▲ The fluid reservoirs and master cylinders looking handsome mounted on the bulkhead. *(Chris Hill)*

▼ Keeping the pipe routing neat and orderly makes a big difference to the overall professional look of the car build. *(Master Mechanic)*

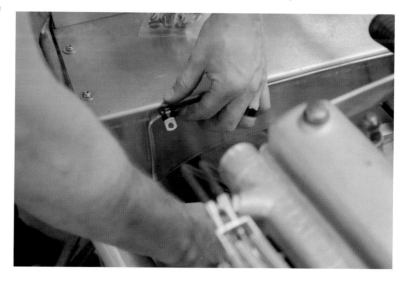

▲ Rubber-lined P-clips like these are perfect for holding fluid lines tight, while isolating them from the worst vibrations. *(Chris Hill)*

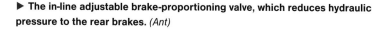

▶ The in-line adjustable brake-proportioning valve, which reduces hydraulic pressure to the rear brakes. *(Ant)*

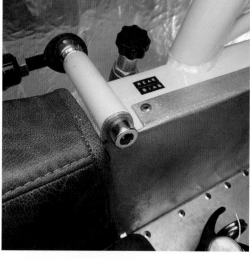

▶ ➕ ▶ ▶ Having access to the rear brake proportioning valve and the pedal bias valve from the comfort of my seat is a nice detail in a race car. *(Ant)*

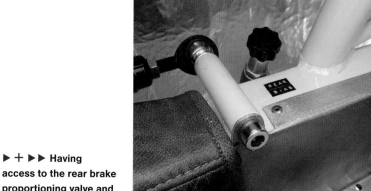

▶ A flexible hose was needed to run to the rear axle. I connected this, and the brake pressure-sensor switch shown here, to a T-piece on the rear crossmember. *(Ant)*

I terminated the line just short of the upper-front trailing-arm mount, so I could add an in-line brake-proportioning valve. This is an adjustable valve, which reduces hydraulic pressure to the rear brakes. This allowed me to set up the system for optimal front–rear performance. This position was perfect for me, as the valve was in a safe space, but at the same time positioned within reach of the cockpit, should I ever wish to use it.

From the brake-proportioning valve, I ran another pipe along the outside of the chassis rail and curved it under to the rear new crossmember. Here, I terminated the line and added a 'T'-piece (using the correct flared ends and fittings). This T-piece was to connect to a brake pressure-sensor switch and a 12in flexible hose running to the rear axle (I temporarily zip-tied the other end of the flexible hose to the axle). The flexible hose was used to

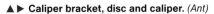

▲ ▶ **Caliper bracket, disc and caliper.** *(Ant)*

accommodate the up-and-down movement of the rear axle. The axle end of the flexible hose has two outlets to allow it to feed both rear brake calipers.

With the rear brake lines fitted from the master reservoir to the rear axle, it was time to fit the rear brake assemblies.

As already mentioned, the rear axle was tailor-made, and I chose to fit Ford bearing carriers to the outer ends to allow a greater choice of aftermarket brake options. I went for a serious spec for the rear, with four-piston aluminium calipers and drilled discs.

Fitting the kit I chose was so simple, it took literally minutes. Once the rear brake assemblies were fitted, I plumbed metal pipes from the flexible hose to each caliper, again using the correct flared ends and fittings. I used zip ties to secure the pipes to the axle.

Knowing that I wanted to fit wire wheels, I finished each rear hub off with a splined hub adaptor. These adaptors suited the spacing of the four wheel-mounting studs on my Alfa Spider axle, and the internal splines of my new wire wheels.

The hub adaptors are handed, so that the wheel-securing spinner threads onto it with a left-hand thread on the right-hand side of the car, and a right-hand thread on the left-hand side of the car. It is essential that the correct adaptor is fitted to the correct side of the car. They are marked, but I double-checked by winding the spinners on. The different threads ensure the spinners are not at risk of loosening when the car wheels are turning forwards.

Front brakes

With the rear brakes plumbed, I moved on to the front. Starting again from the master cylinder, I fitted a copper/nickel line on the outside of the left-hand

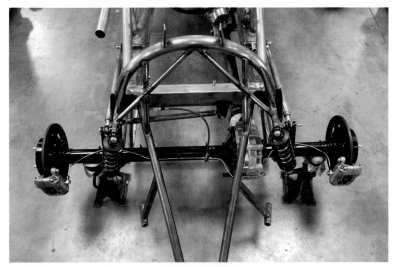

▲ The rear end fully plumbed and with brakes and suspension fitted. Finishing systems completely like this is a very rewarding experience. *(Chris Hill)*

◀ The rear discs are incredibly light. The old drivers like Fangio and Farina would have found them very interesting I'm sure. *(Chris Hill)*

▶ **Hydraulic master-cylinder connections. Low-pressure fluid feed from reservoirs in (braided hoses), high-pressure fluid feed into the brake lines out (metal pipes). The third master cylinder operates the clutch.** *(Chris Hill)*

▶ **I sourced a new remote brake vaccum servo (fitted to most road cars) to provide extra braking power.** *(Moss Europe)*

▶ **The servo in position in the car. The final unit I used had a metal finish, rather than the black-painted finish on the assembly shown above. It was a tight fit behind the radiator...** *(Ant)*

▶ **...as you can see here, with the dry-sump oil tank mounted behind the servo.** *(Ant)*

chassis rail. I made sure that I left a good length of pipe free at the front of the chassis (which would later connect to the brake servo) while I turned my attention to the servo.

I decided to fit a remote vacuum brake servo – this is old-school technology, but very reliable and effective. The engine generates vacuum pressure in its inlet manifold as it runs, and this can be put to good use to assist the driver's braking effort.

The servo comprises a drum with a rubber diaphragm inside. The vacuum created when the engine is running is fed to two chambers in the servo, with the diaphragm positioned between the two chambers. The diaphragm is attached to a piston, and a return spring keeps the diaphragm in the 'neutral' position. When the brake pedal is pressed, an air valve is opened allowing atmospheric pressure to flow to one side of the diaphragm, pushing the diaphragm towards the chamber containing the vacuum, and so pushing the piston. The piston is linked to the master-cylinder piston, forcing the brake fluid into the circuit with more pressure than the driver could generate alone by pressing the brake pedal.

It was important to make sure that the servo was correctly orientated for it to operate correctly. It just so happened that the MG TD front crossmember rose at an angle that suited my needs, and on top of that, the bracket for the MG anti-roll bar provided a great place to position the lower servo stud. For the upper stud, I made a simple steel bracket, with a bend, and welded it to the chassis, on the right-hand side, between the radiator and the front of the engine.

I then plumbed the brake pipe that I previously fitted from the master cylinder to the front of the chassis into the side port on the servo cylinder, and secured that pipe to the chassis with P-clips.

From the servo outlet, at the front of the cylinder, I ran a short fixed line to the centre of the front crossmember and fitted a T-piece. From that T-piece I fitted brake pipes to the existing MG TD brake hose tabs on the chassis.

Again, I went for an aftermarket disc-brake kit for the front end of the car. There is already an off-the-shelf, 'bolt-on' kit out there to convert the 1950s MG TD from drums to discs. I have driven many TDs, and I have to say it's a nice upgrade and very worthwhile – the TD brakes can be somewhat 'wooly'.

As supplied, the aftermarket MG TD bolt-on disc-brake upgrade had the original MG TD five-stud wheel PCD pattern on the hub. Knowing that the front track of the MG TD was narrower than the desired (and wider) 158 front track, I wanted to increase the front track using wheel spacers, just as I had increased the rear track. I knew I could not source a suitable splined wire-wheel adaptor with a diameter big enough to take the original MG

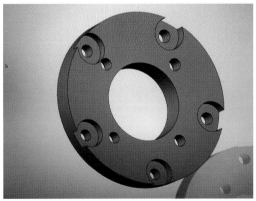

◄◄ The specification for the hub adaptors was first worked out using CAD... *(Paul Cameron)*

◄ ...then machined from a solid billet of aluminium. *(Ant)*

TD stud pattern (which is huge), so I had to get creative, and make my own spacers/adaptors that increased the track and also accommodated a bolt-on splined hub adaptor.

From a solid billet of aluminium, I made adaptors that bolted securely to the MG five-stud hubs, and at the same time provided the Alfa Spider four-stud wheel mounting (to match the rear). These adaptors not only allowed the five- to four-stud conversion, but also acted as spacers to increase the front track to match the 158. Bingo!

With the bolt-on disc-brake assemblies fitted, I fitted the finished wire-wheel splined adaptors to the front hubs.

Finishing touches

With the hydraulic system and brake assemblies completed, I checked each hydraulic-fluid pipe/hose joint, double-checked and carefully tightened.

As this hydraulic set-up used three independent fluid reservoirs, each circuit (clutch, front brakes, rear brakes) needed to be individually bled. Hydraulic fluid is incompressible, which means that any force applied at one end of a hydraulic line is transferred to the other end with great efficiency. If any air is present in the system, because air is compressible, any air bubbles in the system are squashed when the pedal is pressed, leading to a spongy pedal, and reduced efficiency.

Having spent so much time examining the original Alfa 158, I fell in love with the look of the huge brake drums! As my build evolved, I noticed there was a big space between the bodyshell and the rear wheels – it looked odd, so I set about building some 'dummy' brake drums to fill the space. I laser-cut the components for the dummy drums from aluminium, then built them up like a gateau, securing the components together with 6mm bolts. It was quite possibly my favourite part of the entire build!

▶ I used two caliper mounting bolts and a clamp hidden within the dummy brake-drum assemblies to secure them. They looked the part! *(Ant)*

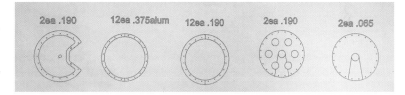

◄ An all-in-one solution, converting the MG five-stud hubs to accept a four-stud wheel mounting, while acting as spacers to increase the track to match the original 158. *(Ant)*

▼ The quantity and thickness of material shown above each layer of my dummy-brake-drum cake!

(Paul Cameron)

2ea .190   12ea .375alum   12ea .190   2ea .190   2ea .065

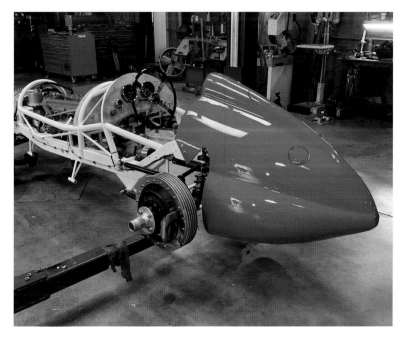

CHAPTER 20

Fuel system

▼ **Electric fuel pumps are compact and simple to fit. Two wires and a basic clamp is all that's needed.** *(Webcon UK Ltd)*

My Alfa Spider donor car left the factory with Spica mechanical fuel injection. The fuel-injection unit is like a little engine in itself, with pistons running in cylinders injecting fuel into the engine at just the right time. The unit can be quite complicated to tune, and although it does perform well when set up correctly, I chose to use a pair of 45 DCOE Weber carburettors on my special. For a start, we needed that induction roar that only Webers can create, but also, it's an Italian-inspired period-looking race car, so it has to run on Webers!

Carburettors and inlet manifold

The kit of parts from Webcon (Weber UK) contained:

- 2 x Weber 45DCOE four-progression-hole carburettors
- 1 x Webcon twin-cable throttle linkage
- 2 x Webcon soft-mount kits
- 4 x Webcon mesh air filters
- 1 x Webcon low-pressure electric fuel pump

I retained the existing water-cooled inlet manifold that came off the Spider. First, I gave it a good clean in my parts washer, and then subjected it to a blasting inside my bead-blast cabinet. The Webers did not

◀ **I am trying to explain to the viewers the science behind matching the inlet manifold ports to the ports on my engine and the carbs.** *(Master Mechanic)*

bolt directly to the existing mountings provided on the manifold, so I ordered some conversion sandwich plates that bolted directly to the Spider manifold and accepted the Webers.

Using new gaskets, I fitted the inlet manifold to the engine, and then fitted the sandwich plates and the Webers to the manifold.

Next, I fitted the linkage kit supplied by Weber to the top of the carbs. I then fitted a pair of accelerator cables to the linkage, drilled two 3mm holes in the bulkhead opposite the accelerator pedal, and passed the two cables through the metal block at the top of the throttle-pedal shaft. I secured the cables by fitting a barrel clamp to the cable inners at the rear of the block. I then checked to make sure that the pedal, cable and linkage all moved freely.

I also removed the long inlet trumpets that came in the Weber kit, as they protrude outside the body line. As a precaution, I covered the inlets with masking tape to prevent any debris getting inside the carbs.

Working from the front of the car backwards I ran a ⅜in braided fuel hose between the two carbs, retaining the 90° elbow on the rear carb and replacing the elbow on the front carb with a 'T'-piece. I used some stainless jubilee clips with chrome shrouds to keep it all tidy.

Fuel filter and fuel pump

I fitted another length of the ⅜in braided fuel hose from the 'T'-piece on the front carb, with a smooth loop down to the chassis rail, running rearwards. I secured the hose to the chassis rail with 'P'-clips as I went.

I opted for a Filter King combined fuel-pressure regulator and filter, with a built-in gauge to enable monitoring of the fuel pressure.

I cut the bracket supplied in the Filter King kit to size, and welded it to the chassis rail above the gearbox. I then connected the fuel line to the filter assembly, again P-clipping the fuel line to the chassis as I went.

◄ **The throttle linkage for the Webers. There is an unmistakable sound when these carbs are opened up on a long straight – like music.** *(Ant)*

◄ **There are two throttle cables, that pass through the bulkhead and attach to the back of the pedal.** *(Ant)*

◄ **The space available to connect the cables to the pedal was tight. Once they were connected, I checked for free movement to make sure they did not catch anything.** *(Ant)*

◄◄ **I used a Filter King fuel filter and pressure gauge in one unit.** *(Master Mechanic)*

◄ **The unit attached with two bolts. The supplied bracket needed cutting and welding to the chassis.** *(Ant)*

▶ **I welded a steel plate to the chassis, then drilled and tapped an 8mm hole to mount the fuel pump.** *(Ant)*

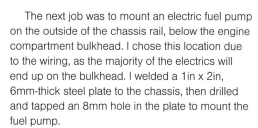

The next job was to mount an electric fuel pump on the outside of the chassis rail, below the engine compartment bulkhead. I chose this location due to the wiring, as the majority of the electrics will end up on the bulkhead. I welded a 1in x 2in, 6mm-thick steel plate to the chassis, then drilled and tapped an 8mm hole in the plate to mount the fuel pump.

I then bolted the fuel pump to the plate, and connected the fuel line from the Filter King to the pump. The fuel pump itself is directional, so I had to make sure it was fitted the correct way round, taking note of the flow-direction arrow.

From the rear of the pump, I ran a fuel line back towards the fuel tank location (above the rear axle), leaving sufficient spare hose to connect to the tank. I P-clipped every 8in as I went, and used zip ties when I reached the rear, round chassis tubes.

Fuel tank

There were a number of options for the fuel tank. There are so many available on the market that it

▶ **The fuel pump, with attached in-line filter, plumbed in and bolted to the bracket.** *(Chris Hill)*

▶▶ **I riveted aluminium P-clips, with rubber inserts, to the chassis to secure the fuel lines.** *(Chris Hill)*

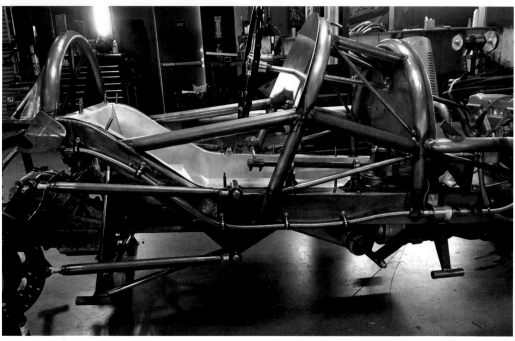

▶ **The fuel line clipped neatly in place.** *(Chris Hill)*

◄◄ **I wanted to fill the space available at the rear with as big a fuel tank as possible.** *(Ant)*

◄ **I could have simply bought a tank ready to go, but I would have had to compromise on the capacity, and it would not have looked as cool!** *(Ant)*

would have been easy to simply select a tank that fitted within the tail section of the car and then work out a way to secure it. That's exactly what I was going to do, until one Sunday, when I was itching to get into the workshop – so I went in and made myself a tank.

I already knew the dimensions of the tail section, as the triangular rear chassis and tail-support tubing dictated it. This was the space in which I was going to build the tank.

Using some 2.5mm aluminium sheet, I first cut the front tank wall, with a return to the top and two

folds to shape it around the axle, allowing room for the movement of the axle. I clamped the sheet in place on the chassis as I went.

I then made a triangular top panel and rolled it through my sheet-metal rollers to shape it. I made a similar triangular bottom panel that met the bottom edge of the top panel (all within the rear chassis support). Next, I clamped the panels and tacked them in place using a TIG welder.

Removing the three-sided assembly from the chassis, I placed it on the floor, laid a piece of aluminium underneath, and drew around it to mark

◄◄ **Aluminium sheet is very easy to work with, but having access to a sheet-metal folder helps tremendously.** *(Ant)*

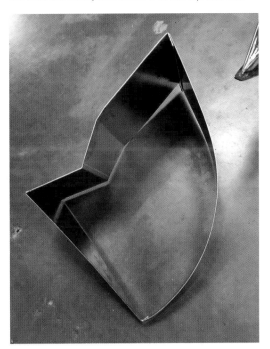

◄ **Trial-fitting the tank before adding the finishing touches.** *(Ant)*

▲ The filler-cap opening in the body was cut during the bodyshell build. I marked through the opening where the filler neck sat. *(Ant)*

▲ I drilled holes, then welded threaded bushes in place for fuel outlet and return lines. *(Ant)*

▲ With the tank fully welded, I folded two aluminium strips to create long brackets, and welded them to the tank sides. *(Ant/Paul Cameron)*

▲ I also fitted a vent with a safety rollover valve to the top of the tank. *(Ant)*

▶ Special fuel foam stuffed in through the filler hole prevents the fuel from sloshing around too much. *(Master Mechanic)*

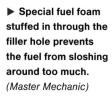

▶ I fitted short trumpets to the Webers. The ones originally supplied were too long and fell outside the body line. *(Master Mechanic)*

out a side panel. I then turned the assembly over and repeated the procedure to mark out the other side. Next, I cut the side panels from the marked sheet, tack-welded them in place, and rested the unit back in the chassis. I tacked it all in place. At this stage, the only remaining pieces were the lower side panels. I made a quick check that the tank still fitted, and then cut and fitted the lower side panels.

I then refitted the body tail section, and placed an off-the-shelf aluminium fue filler neck and cap in the opening of my fuel filler flap, which was cut during the bodyshell build. I leaned in and marked the location of the filler neck on the top of the tank with a pen.

It was now time to remove the tail section again, and fold two sections of 90°-angled 1in x 1in aluminium, resting the aluminium sections over the two rear chassis tubes, before tack-welding them to the tank sides to form mounting brackets.

I stood back to admire the shape and fit, and checked that the axle didn't foul the tank when the suspension was compressed. I then removed the tank and set about fully TIG-welding it. At this stage, I drilled a small hole at the bottom-right front corner of the tank, where the fuel outlet was going to go, and carried out a quick pressure test, using an airline to check that the tank was sealed, and that my welds were airtight.

I then welded a threaded bush to the hole I had drilled previously for the fuel outlet, and screwed a ⅜in brass outlet into the bush. I also drilled a second hole above the first, and added a threaded bush for a return line (in case I needed to add one later), which I blanked off using a screw-in plug at this stage. I used PTFE tape on the threads to ensure fuel-tight seals on both the fuel outlet and the plug.

I also drilled a hole on the top of the tank, in front of the filler neck, and added a threaded bush to house a vent with a safety rollover valve.

I placed the tank back into the chassis for the last time, then drilled mounting holes and tapped threads in the chassis, before bolting the tank in place. I then connected the fuel line from the pump to the tank outlet, and carefully P-clipped and zip-tied the remaining length of line to the chassis.

As this is a track car, I need to avoid the fuel sloshing around inside the tank under cornering, braking and acceleration (which could cause fuel-feed problems), so I cut some blocks of specialist fuel foam using a band saw and shoved them down the filler neck!

Finally, I returned to the Webers, and fitted shorter trumpets that cleared the body. Once fitted, I protected the open trumpets with masking tape again to avoid any debris entering the carbs.

Fuel system completed!

CHAPTER 21

Cooling system

The ethos for a lot of this build was to keep things clean and simple, and that's exactly what I did with the cooling system.

The main thing to consider was fluid level. The system has an inherent 'water table' – that is the level at which the coolant naturally sits in the engine. It was important to consider how air could get trapped, and therefore how to fill and bleed the system, and also how to drain the system.

I spent hours online examining numerous radiator dimensions, and was just about to make one from scratch, when I found an off-the-shelf radiator to fit a 1932 Ford Model B 'Tudor', manufactured in aluminium. Not only did this radiator fit within the nose shape of my car, it had a cool curved top that complemented it too.

Most importantly, the top of the radiator, when installed in the car, was higher than the thermostat on the top of the engine. This meant the filler cap was the highest point of the 'water table', which would be great for filling the system and bleeding the air. With many regular cars the radiator is below the highest point of the engine, as the nose dips down, so using the filler cap

◄ **I was about to build a radiator from scratch, when at the 11th hour I found this one, made in aluminium for a 1930s Ford Model B.** *(Champion Cooling Systems)*

to bleed trapped air is often not possible. The classic bullet-shaped front of my car helped me tremendously here.

Sadly, the radiator wasn't going to be a simple bolt-on job – it would require some small modifications to work in my car.

I decided to place the radiator as far forward as possible, in front of the steering rack. Like the real 158, it would lean back very slightly to fit.

First, I cut the two existing upper brackets off both sides of the radiator. I kept the lower brackets, as they would work well for my car. The Ford radiator fitted perfectly between the front chassis 'horns' of the MG TD – as if it was tailor-made!

I made a pair of small 1in x 90°-angled brackets, both 2in long, and tack-welded them to each side of the radiator while it was clamped to the MG TD chassis horns. I then drilled through the brackets into the chassis, and bolted the brackets in place with a pair of 6mm bolts and nuts.

Again working on each side, I then cut a 6in piece of 1in flat bar (4mm thick) to secure each radiator lower bracket to the bottom of the chassis. I tack-welded the bar to the bottom face of the chassis, and then drilled through the lower end of the bar into the radiator bracket, again fixing each side using a bolt and nut. The radiator was now secured to the car.

I double-checked the radiator position with the bodyshell fitted, to ensure that it fitted within the nose. The body shape dives down quite low at the front so I didn't want to rely on the fact that the top of the radiator was lower than the front chassis hoop. I then removed the radiator, and fully

▲ **The original Alfa 158 had a radiator that leans back, so it was a nice touch that mine does too.** (Chris Hill)

▶ **The lower brackets worked just fine as they came – a simple case of drilling and bolting in the right place.** (Chris Hill)

▶▶ **The top brackets needed a bit more work. I also had to double check that the body fitted OK with the radiator at this angle.** (Chris Hill)

welded the brackets in place on the radiator and the chassis.

The Ford radiator has two pipe stubs at the top – one for the top-hose connection and one for the filler neck. I cut them both off, and welded a cap over the hole for the top-hose connection.

I kept the outer portion of the filler neck, and welded an extra length of tube to it. I then drilled a hole in the top centre of the radiator to locate the modified filler neck.

I refitted the radiator again, and tack-welded the new filler neck in place, ensuring that it fitted under the front chassis hoop, and that it was positioned

◄ This is how the Ford radiator looked before 'modifications'. It's unlikely that 'standard' items are going to fit first time, and I had to get creative. *(Chris Hill)*

◄◄ I completely sealed the existing top-hose hole and cut off the original filler neck. *(Ant)*

◄ I lengthened the filler neck, retaining the original cap and housing. *(Ant)*

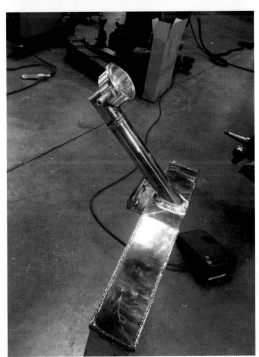

◄◄ Here, the modified radiator is in place for a trial fitting. The lengthened filler allows me to fill the system from under the bonnet, and the filler is still the highest point in the cooling system. *(Ant)*

◄ Fully welded, the assembly is now ready to install in the car. *(Ant)*

► The radiator top hose can be seen here, connecting the thermostat housing on the engine to the radiator. *(Ant)*

▼ The bottom hose runs under the chassis. I used existing hose sections sleeved together with straight pipe to provide the hose run I needed. *(Ant)*

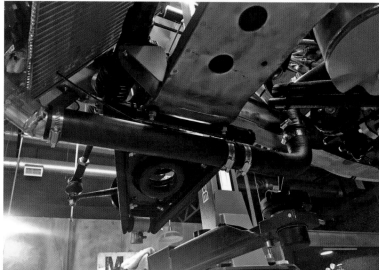

► I'm proud of this cute little saddle bracket which supports the bottom hose. It's the little things that give the greatest pleasure, I find. *(Ant)*

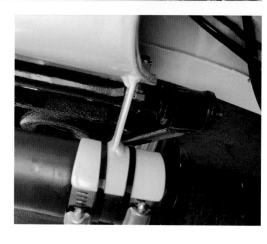

parallel to the floor. This allows the radiator to be hidden within the nose, but allows easy access to the filler neck within the engine bay with the bonnet removed. It looks neat and tidy too.

I then removed the radiator from the chassis once more, and fully welded the filler neck in place.

Next, I cut another piece of tube to form a new radiator top-hose outlet to fit to the hole where the filler neck used to be. I angled the outlet slightly towards the thermostat housing on the inlet manifold, and welded it in place. Once the components had cooled, I ran a flexible hose (actually a Land Rover part) from the outlet to the thermostat housing, and secured it in place with jubilee clips.

I retained the existing radiator lower outlet, but welded a two-piece extension tube to it, so that the end of the extension sat parallel to the floor, and the bottom hose would clear the chassis crossmember. I made sure that the diameter of the extension tube matched the bottom-hose diameter. The next task was to use rubber hosing to connect the radiator bottom outlet to the water pump on the front of the engine. I did this using two 90° sections (sleeved together with an internal tube) and a straight section to connect to the radiator, positioning the hoses to ensure sufficient clearance around the surrounding components. I later welded a 'saddle' bracket to the chassis to support the lower hose.

To ensure sufficient cooling, I elected to fit an electric fan, and went big, with a 14in-diameter one! I decided to fit a fan-override switch on the dash, so that I could manually switch on the fan from the cockpit (just in case).

I fitted the fan onto the front of the radiator core, using purpose-made radiator-fan tie straps (supplied with the fan).

That was it – cooling system complete!

▼ It's best to find the biggest-possible fan that will fit in the space available. This one is 14in diameter. *(Master Mechanic)*

CHAPTER 22

Engine ancillaries

The original Alfa 158 engine bay was designed for a straight-eight engine, so my tiny four-pot seemed quite small! The good news was that there was plenty (I say 'plenty' with my tongue firmly in my cheek) of space to fit the various other components around the engine. The trick was to make the best possible use of that space, and carefully planning the location of each item in advance was essential, as was measuring for clearances, etc.

Lubrication system

The original Alfa 158 featured a dry sump, a system found on many racing cars. A dry-sump system, like a conventional oil system, is designed to supply oil

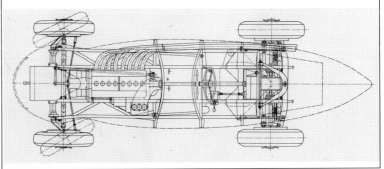

▲ **An original factory drawing of the Alfa 159. See how it compares with my car above.** *(Centro Documentazione Alfa Romeo – Arese)*

▲ The dry-sump oil pump is powered by the same belt as the original fuel-injection pump. *(Ant)*

▼ A view of the dry-sump pump from underneath the chassis, with the engine installed. *(Chris Hill)*

to the critical engine components, but it has a few significant advantages over a regular lubrication system where oil is kept sloshing around in a 'bath' (sump) at the bottom of the engine. The advantages of a dry-sump system are:

- The oil is kept in a remote tank separated from the engine, so it is inherently much cooler than it would be sitting in a conventional sump.
- The separate tank allows an increased oil capacity.
- The crankshaft is not sitting in the bath of oil (in the sump), which actually causes drag and reduces efficiency.
- With a conventional sump system, high-speed cornering can lead to oil starvation as the oil sloshes from side to side within the sump. A dry sump eliminates this.
- The engine can be mounted lower in the chassis, as a dry-sump pan is far shallower than a conventional wet sump (attached to the bottom of the engine).
- A dry sump does not have a 'wing' like the original Alfa Romeo Spider sump, so it slots nicely within the MG TD chassis rails.

I opted for an off-the-shelf dry-sump kit that required a little extra engineering to make it work. My kit came with numerous components, including a new shallow sump, oil tank, oil pump, remote oil filter, breather tank and a few other peripheral items.

Dry-sump pump

The dry-sump system relies on a belt-driven pump, driven from the crankshaft via the same sprocket that, on the Spider engine, originally drove the mechanical fuel-injection pump. I made a bracket from ½in aluminium plate that bolted to the cylinder block using the mounting holes originally used for the fuel-injection pump. I reused the original fuel-injection-pump sprocket from the Spider, fitting it to the oil pump, then fitted the oil pump to the bracket, making sure that the sprocket was precisely aligned with the crankshaft sprocket in exactly the same position as when it was fitted to the fuel-injection pump. This meant that the original-spec drivebelt (a toothed belt – absolutely essential for running a dry-sump oil pump) matched the two sprockets. I fitted a new belt, as it was not worth the risk of reusing the old belt – a broken drivebelt means no oil flow, not good!

Before fitting the new, shallower sump pan, I had to modify the existing Spider engine-oil pump, which sits in the bottom of the crankcase. I removed it, cut the bottom off the driveshaft (which also drives the distributor), shortening it to make sure it did not protrude from the bottom of the modified pump, then cut off the bottom of the pump body to leave

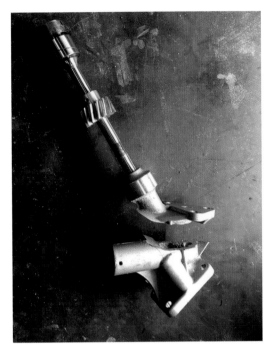

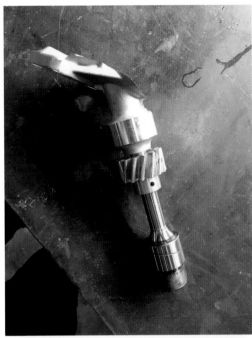

◀◀ The old Spider oil pump fitted to the original sump was now too long. Like a lot of things in this build, it needed modifying. *(Chris Hill)*

◀ The oil-pump driveshaft also drove the distributor, so I couldn't simply remove it. The modified unit still drives the distributor, but the original oil pump is disabled. *(Chris Hill)*

just a thin section around the mounting holes. I plug-welded the pump inlet, and then refitted the pump. The driveshaft continued to drive the distributor, as before, but no longer drove the oil pump to pick up oil from the sump.

Next, I fitted the new sump pan, using a new gasket.

Remote oil filter

Engine oil needs to be filtered to prevent particles of debris from damaging the fast-moving engine components, particularly the bearings. A regular car oil filter will trap debris that is around 20 microns in diameter – that's 0.02mm – which is

◀ I cut the bottom off the original oil pump and plug-welded the pump inlet before refitting the modified assembly. *(Chris Hill)*

◀◀ The modified original oil pump in place in the crankcase, ready for the new dry-sump pan to be fitted. *(Ant)*

◀ The dry-sump pan is very shallow and neat, and it provides extra ground clearance. *(Ant)*

▶ I fitted a remote oil filter unit on the inside edge of the left-hand chassis rail, below the hydraulic fluid master cylinders and, importantly, aimed it downwards to avoid spillage when renewing the spin-on filter. *(Ant)*

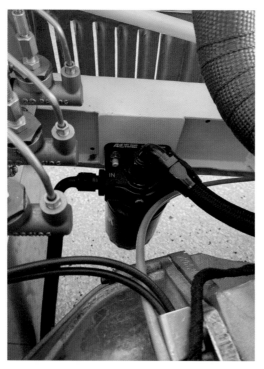

smaller than a human white blood cell. Oil filters are pretty effective!

Because oil filters are so effective at trapping particles, they do need to be renewed at regular intervals. As the filter material traps more dirt, the oil pump has to work harder and harder to push the oil around. The oil circuit has a filter bypass valve, which will open if the filter is clogged – while this will avoid critical engine damage due to oil starvation, it is effectively like not using a filter at all, and there

is no indication that the valve has operated, so you won't know.

I fitted a remote oil filter unit on the inside edge of the left-hand chassis rail, below the hydraulic fluid master cylinders and, importantly, aimed it downwards to avoid spillage when renewing the spin-on filter.

Oil tank

The oil tank was the biggest component (other than the engine!) to be fitted within the engine bay. I chose to fit mine in front of the engine, and mounted it using two brackets welded to the chassis, making use of the tank clamps supplied in the kit.

Oil catch tank

As the engine runs, tiny amounts of oil and fuel sneak past the piston rings and into the crankcase. This can increase crankcase pressure, which isn't good. Some older racing cars used to simply vent this pressurised vapour to the atmosphere, but this isn't healthy for polar bears, and today an 'oil catch tank' is usually required by the regulations before a car is cleared to go out onto the track (or indeed road). I fitted a basic catch tank, sometimes called a breather tank, which condenses the vapours, separating the oily stuff from the air. The oil and fuel mix sits in the tank, and the air vents out.

This catch tank can be mounted anywhere convenient, but it must be mounted higher than the oil tank. I used a bracket to mount mine on the engine compartment bulkhead, with a hose connecting the tank to the crankcase breather.

▶ The two oil-tank clamps attached to the finished brackets on the chassis. *(Chris Hill)*

▶▶ The oil tank in place on the finished car. This took up the last of the available space at the front of the engine bay. *(Chris Hill)*

Alternator

Some race cars run without an alternator. That may seem crazy, but depending on the electrical loads, the battery type, and the length of the event, the charged battery can be enough to run the engine for one race, and the battery can be charged between races. The benefit is that the engine isn't robbed of power as it drives the alternator. I did think about taking this route, but decided that I preferred the sensible route of having a small alternator to charge the battery all the time.

The alternator that came off the Spider was rather large, so I decided to fit a much smaller alternator than the original. There are a few companies that specialise in smaller alternators, some of them offering a high output in a very compact unit. I robbed the pulley from the original alternator, and bought a new drivebelt to suit.

Battery and tray

'Standard' car batteries are a significant weight when considering a race car, so I opted for a small, lightweight but high-capacity race battery. The battery had to be securely mounted, and I chose to fabricate a basic tray from sheet aluminium to sit on the flat panel at the rear of the engine compartment. I used a strap-and-bolt arrangement to secure the battery firmly in place.

The leads pass neatly through the bulkhead, protected by rubber grommets. The black cable attaches directly to one of the gearbox-bellhousing bolts as the main earth cable. The red cable bolts securely to the contact on the main battery cut-off switch.

Fusebox

All the wiring circuits need to be protected by fuses. The fuses are designed to 'blow' and break the relevant circuit if the current through the wiring exceeds the capacity of the fuse. The fuse fitted to each circuit must be of the correct rating. It is useful if the fuses are within easy reach for quick replacement if necessary, so I fitted the fusebox on the bulkhead, next to the battery. Further details are given in Chapter 25.

Ignition coil and distributor

I cleaned and refitted the standard Alfa Spider distributor, but I fitted new points, cap and HT leads.

The role of the ignition coil is to ramp up the battery voltage from 12 volts to around 40,000 to provide HT voltage to create sparks at the spark plugs.

I fitted a new standard ignition coil, as they aren't expensive, and as the engine's horsepower was not significantly increased, a regular unit would do just fine. I noticed a handy little hole at the rear of the engine cam cover that was crying out at me, so I mounted the coil there, using a clamp secured by a nut and bolt.

◄ The oil catch tank in position on the bulkhead. This condenses the oil vapours, separating the liquid oil from the air expelled from the crankcase. *(Chris Hill)*

◄ Some race cars run without an alternator. I opted to fit one, and this compact unit worked perfectly for me. *(Chris Hill)*

◄ I mounted the fusebox next to the battery, with a watertight, removable lid to finish it off. *(Chris Hill)*

◄ I fitted the ignition coil using a handy spare bolt hole at the rear of the cam cover. *(Chris Hill)*

CHAPTER 23

Exhaust

▶ Then-current World Champion Nino Farina adjusts his goggles prior to a test session at Monza, 1 June, 1951. As well as showing the purity of the car's shape, and its exhaust system, this stunning image by *Paris Match* photographer Maurice Jarnoux clearly shows the Magneti Marelli starting equipment.
(Maurice Jarnoux, Getty Images)

On a 1930s single-seater racing car that has a very pure body shape – no trim or fancy chrome – the exhaust becomes an important part of the car's aesthetics, as well as its performance. It's a big, bold feature on the outside of the smooth, minimal, svelte shape.

The original car was called a 158 because it was fitted with a 1.5-litre, eight-cylinder engine. So, obviously, with one exhaust port per cylinder, that meant the car had a glorious bank of eight exhaust branches protruding through the body. The exhaust exited on the right-hand side on the original car.

My car, however, has a more modest powerplant, and as I used the engine from an Alfa Romeo Spider, it meant that I only had a four-cylinder engine with four exhaust ports. Like many manufacturers of production road cars, Alfa Romeo chose to use a cast-iron manifold for the Spider, that exited downwards to take the exhaust gases under the floor of the car and out at the back.

I decided to get creative and build myself a bespoke tubular manifold that would exit through the side of the bonnet and down the outside of the body, just like

▲ **I bought mild-steel tubes with one 180° and one 45° bend.** *(Master Mechanic)*

the real 158. To top it off, I made it look like an eight-cylinder engine to mimic the real car. Four of the tubes are just for show – sneaky! Of course, my exhaust exits on the left-hand side as my engine dictates.

Using some off-the-shelf pre-bent 1¾in 16SWG mild-steel tubes, with both 180° and 45° bends, I started at the front exhaust port on the engine and worked my way backwards. I tackled the manifold in two halves. First of all, I built the part of the exhaust that exits the body – the section that you can see on the outside.

I built the exhaust on a flat, steel workbench for ease and accuracy. For the front four stubs I didn't need to worry about exhaust flow, I was merely welding four 90° bends together to give the appearance of an extra four cylinders feeding into the manifold.

I started by cutting at the centre of the 180° bend to give two 90° sections. Holding one end of the 90° bend in my chop-saw with the blade at 90° to the vice actually made the cuts quite accurate, with minimal shaping required with a grinder to mate the tubes together. I made the tail on the first 90° bend 12in long – enough to accommodate the three remaining front stubs.

Once I had cut and shaped the first four tubes, I used a 1¼in solid-aluminium spacer to hold between each tube to ensure consistent and accurate spacing, then I tack-welded all the bends together, tack-welding the three shorter tubes to the tail of the first tube.

I then fully welded each joint. This front section was purely about aesthetics, and had nothing to do with exhaust gas flow, function or performance. The rest of the exhaust, of course, needed to actually work!

▲ **My first job was to cut the 180° bend in half. You only need a few bends like this to make all manner of shapes.** *(Master Mechanic)*

◀ **Exhaust tubes like this cut really easily. I used the saw shown here.** *(Chris Hill)*

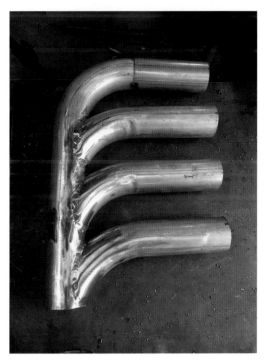

◀ **It takes a lot of time to weld up pipes like this, and grind them down smooth, but I feel the result is more than worth the effort.** *(Chris Hill)*

▶ **The four rear stubs ready for welding to the straight tube.** *(Chris Hill)*

▶▶ **The idea is to not impede the exhaust gas on the way out. Great performance gains can be made by optimising exhaust flow.**
(Ant/Paul Cameron)

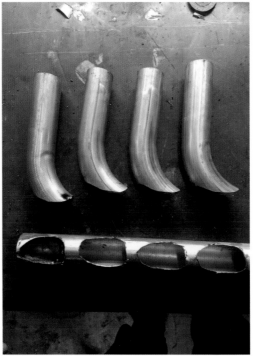

▼ **The front half of my manifold was just for show, to mimic the original car. The rear half had to fit to the engine perfectly, and I had to block off the front section of the straight tube to prevent exhaust gases from flowing forward instead of out through the tailpipe.**
(Chris Hill/Paul Cameron)

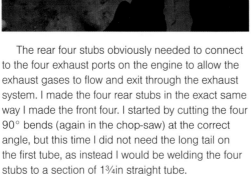

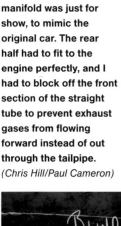

The rear four stubs obviously needed to connect to the four exhaust ports on the engine to allow the exhaust gases to flow and exit through the exhaust system. I made the four rear stubs in the exact same way I made the front four. I started by cutting the four 90° bends (again in the chop-saw) at the correct angle, but this time I did not need the long tail on the first tube, as instead I would be welding the four stubs to a section of 1¾in straight tube.

Again, I used the same aluminium spacer to give the correct spacing between the tubes, and I marked around the oval cut end of each tube on the straight tube with a pen. I then cut out the four sections I had

marked for the ends of the tubes using an angle grinder, and finished off the holes with a small die grinder. I spent time matching the holes as closely as possible to allow for maximum exhaust flow.

It was essential that at the point where the first tube entered the straight pipe, the pipe was modified internally to direct the exhaust gases to flow rearwards along the pipe. So, I cut a small scallop of leftover bent tube, small enough to be inserted inside the front section of the straight tube, and welded it in place. This ensured that the exhaust gases flowed from the first port back towards the rear of the car. If I hadn't completely blocked off the straight pipe forward of this first port, some of the exhaust gases would have been able to escape up the front of the pipe into the 'fake' forward section. Using an off-cut of the pre-bent tube allowed me to create a smooth and flowing passage for the gases, essentially continuing the contour of the 90° curve, but inside the straight tube.

Again, using the aluminium spacer between each stub to ensure correct spacing, one by one I welded the four stubs to the straight tube, ensuring that the ends of the tubes aligned with the holes.

I then welded the front section, with the four 'dummy' stubs, to the rear section, to create the eight-branch manifold.

It didn't matter at this stage that the tubes were of different lengths, as they needed to be trimmed later. What mattered was that they were all evenly spaced and the bends flowed and curved uniformly. The finished exhaust would become an important part of the aesthetics of the car, so it had to be accurate.

This was simply one half of the manifold (the

◀ **The almost-finished manifold awaiting the final cutting of the pipes to length.** *(Chris Hill)*

▲ **Exhausting work! The TV cameras seem to love it every time I create sparks.** *(Master Mechanic)*

external half). To build the internal half of the manifold, I first refitted the front section of the body shell to the chassis. I bolted the bodyshell to the chassis to ensure that it sat in the correct position and didn't move.

The complete eight-branch manifold is quite large, and so is the bonnet opening on the bodyshell. Using a tape, I measured both the manifold and the bonnet, and then sat the manifold in the centre of the engine-bay opening. I marked ¾in clearance on the body, front and rear, either side of the manifold. I also decided to trim the manifold tubes at this stage, long enough that the stubs were inside the bonnet, but allowing for a reasonable curve in the internal manifold which would be joined to them.

Next, just to recheck the ¾in clearance at each end, I hovered the manifold between these desired location marks, with the inside edge of the straight exhaust exit pipe sitting perfectly in line with the line of the bodyshell running towards the cockpit. I also needed to leave ¾in clearance at the top and bottom of the manifold, so bearing in mind that my exhaust tube diameter was 1¾in, I measured down from the top edge of the bodyshell by half the diameter of the tube (⅞in) and added the ¾in clearance to give 1⅝in, then drew a horizontal line this distance from the top of the bodyshell, between the two marks I made previously at the front and rear.

I made a final check, then cut the body using an air saw so that the manifold would eventually sit half in the bodyshell and half in the bonnet. I cut a radius at each end of the slot to match the radius of the exhaust tubing.

◀ **Cutting the bodyshell to accommodate the manifold. As always, I measured twice and cut once.** *(Master Mechanic)*

▼ **The bodyshell and bonnet trimmed to give clearance around the manifold.** *(Chris Hill)*

▶ **I held the manifold in place to check the clearance once I had cut the slots in the bodyshell and bonnet.** *(Chris Hill)*

▶▶ **Holding the manifold in place here shows that I had a challenge ahead to fabricate a manifold to join the exhaust system to the ports on the cylinder head.** *(Master Mechanic)*

▶▶ **Using the Alfa Spider cast manifold as a template, I made a new manifold flange, shown here secured by the manifold nuts.** *(Master Mechanic)*

▼ **It's enough to send you round the bend! Here, I'm holding one of the 'S'-shaped tubes to join the exhaust to the flange on the cylinder head.** *(Master Mechanic)*

Once the slot was trimmed, I temporarily tack-welded the manifold to the side chassis rail, using some offcuts of tube, to hold it securely in place in its correct final position while I set about working on the internal section.

I then hovered the bonnet over the bodyshell, marked it in the same way as I marked the edge of the bodyshell previously, and cut a corresponding slot. Next, I fitted the bonnet, and making small adjustments, I tickled the exit slot in the bodyshell and bonnet so that it was uniform and straight, before double-checking that I was happy with its final shape. How cool was this going to look?

I removed the bonnet again, which revealed the scale of the challenge ahead. The plan was to join the rear four manifold tubes to the exhaust ports on the engine, while at the same time working around the routing of the steering-column run from the bulkhead to the steering rack. Easy!

Using the old Alfa Spider cast-iron manifold as a template, I made a new manifold flange from some 8mm mild-steel plate. Another option would have been to have this laser-cut to save time, but I decided to make it myself.

I protected the engine exhaust ports with masking tape to prevent anything entering the engine, and I then fitted the 8mm flange to the cylinder head exhaust-manifold studs and tightened the nuts. Next, I had to join the steel flange to the external part of the manifold with smooth, flowing tube shapes.

The Spider manifold branches are angled downwards to direct the exhaust flow into the exhaust system under the car. This was actually rather helpful, as it directed the first exit tube down and away from the steering column. The curve that was naturally needed to then smoothly reach the external part of my manifold meant that the

◀ ◀ The first three 'S'-bends held in position with quick-release clamps. The clamps don't like the heat, so care is required when welding near them. *(Ant)*

◀ With the pipes initially welded in position, I refitted the steering shaft to check on clearance. *(Ant)*

▼ The finished, polished manifold. Now for the main pipe. *(Ant)*

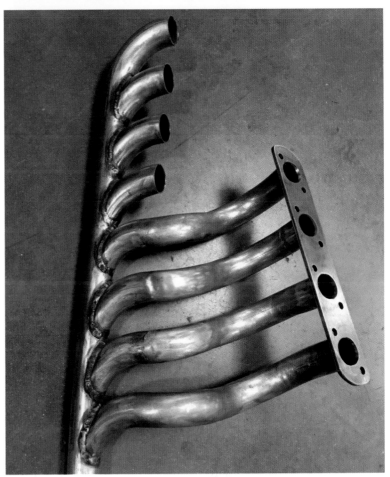

completed manifold would have a 'valley' in it to accommodate the path of the steering column.

Creating nice smooth 'S' shapes from sections of pre-bent steel tube was going to take a bit of time. The number of pieces of tube I used for each section wasn't really important, as they would be fully welded into one flowing shape. Working on one at a time, I tried to match each tube with a degree of symmetry.

Once one tube shape was established, I fully welded each 'S' bend on the workbench, and with a linisher, polished the sections into one smooth pipe. It was far easier to tackle the weld and polish on each individual section, rather than having to tidy the whole finished, welded manifold later.

I then held the individual 'S' shapes in place with C-clamps as I worked my way along one tube at a time. Once happy, I tack-welded them all in place between the 8mm flange on the cylinder head and the external section of the manifold.

Assembling the whole exhaust manifold took some time, but this is exactly the type of fabrication I love the most. All the way through the process I kept a careful eye on the steering column to make absolutely sure that all the components cleared each other.

Once the whole exhaust manifold was finished, I removed it, fully welded it, and then gave it a thorough grind-down and polish with an orbital sander.

With the manifold completed, I moved on to the main exhaust pipe. I wanted the pipe to

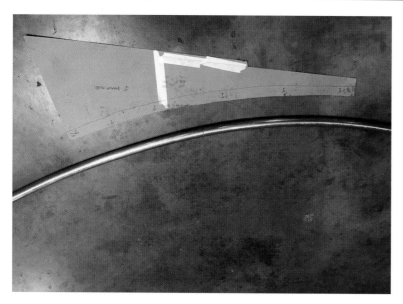

▲ It's possible to bend the main pipe with a lot of little cuts and welding, but I decided to have it rolled by a specialist. *(Chris Hill)*

curve and follow the shape of the bodyshell. I didn't have a set of tube rollers in the workshop, so I made a cardboard template to follow the curvature of the body at the height of the exhaust, and took the template to a steel-fabrication shop, where I had some larger 2in mild-steel tube rolled to match the template.

With the manifold refitted to the engine and bolted in place, I first offered the rolled section up against the body, allowing sufficient clearance bearing in mind the heat and, of course, clearance around the rear suspension. I positioned the main pipe so that it met the end of the manifold, and checked the fit around the body.

I then removed the main pipe and the manifold again, and using an exhaust-flaring tool, I flared open the 1¾in manifold tube to 2in, trimmed the manifold tube and the main pipe to fit end-to-end, and again refitted the manifold to the car.

I used a butt weld to hold the front pipe in place on the manifold, and placed a chair to temporarily support the rear end, then made an exhaust mount from the clamp part of an off-the-shelf 2in exhaust

◄ Following the cardboard template worked perfectly, and the finished pipe fitted the body really well. *(Ant)*

▼ I adapted an off-the-shelf exhaust bracket, fitting 'cotton-reel' rubbers for vibration protection. *(Ant)*

◄◄ I fitted the mounting bracket to the body, then fitted the exhaust, to work out the exact position to weld the metal bracket to the pipe. *(Ant)*

◄ The bracket welded to the exhaust. It will be coated with the rest of the exhaust system. *(Ant)*

U-clamp. I discarded the U-bolt, and fitted two 'cotton-reel' rubber exhaust mounts to the clamp, securing them with nuts.

I chose a position behind the rear-suspension opening in the bodywork for the mounting, then drilled two holes to secure the rubber cotton-reel mounts through the body. I tack-welded the clamp directly to the exhaust pipe, and later neatly TIG-welded it in place.

To complete the authentic look, I then cut the rear of the pipe at a 'slashed' angle for that aggressive period race-car finish.

I made one final check, and it all looked fantastic. Of course, I fitted a wheel and tyre on the rear axle so I could get a better sense of how it flowed. I was thrilled! I then removed the (now) one-piece exhaust/manifold and fully welded it, tidied the welds and polished it up. I decided not to fit a silencer at this stage, as I wanted to listen to how she sounded. I was after that deep sound those epic grand prix cars are famous for. However, I had to bear in mind that some race tracks have a decibel limit that would need to be considered.

With the completed exhaust system removed, I raised the car on the ramp, and reaching up from the underside of the car, I made a 4in x 4in flat steel plate from 4mm mild steel, and marked the position of the two body exhaust mount holes on it. I then temporarily bolted the plate to the body through the exhaust mounting holes, and using some ¾in round tube, I added triangulation to the chassis tube on which the fuel tank was mounted, to provide

► Finally, I cut the rear of the pipe at a 'slashed' angle to complete the authentic look. *(Ant)*

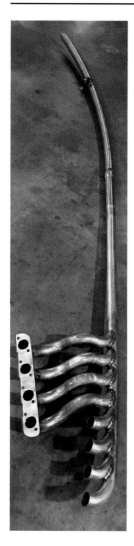

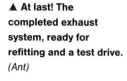

▲ At last! The completed exhaust system, ready for refitting and a test drive. *(Ant)*

▲ I made a simple exhaust-mounting bracket triangulated off the chassis and hidden behind the body. *(Ant)*

extra internal support. I then removed the tail for better access and fully welded the bracket and triangulation in place.

I then refitted the complete exhaust system to the car and moved on to the next stage of the build.

In actual fact, I got my car running (minus the body), so that I could hear the exhaust (and check other functions of course).

After this 'test drive' I concluded that the exhaust

▲ The modified off-the-shelf bracket, with 'cotton-reel' rubbers, is shown here welded to the exhuast, and attached to the chassis bracket. *(Ant)*

sound just wasn't right for me, and a silencer needed to be added. I sourced a 17in x 3in 'Cherry Bomb' silencer, and with the bodyshell refitted (to ensure the necessary clearance), I cut a section from the exhaust alongside the cockpit and fitted the silencer in place.

Once I was happy with the final position of the silencer, I tack-welded it in place in the exhaust system and removed the whole system again. I then

▶ The exhaust fitted to the car, without the bodyshell. This photo was taken just before the very first test drive. *(Chris Hill)*

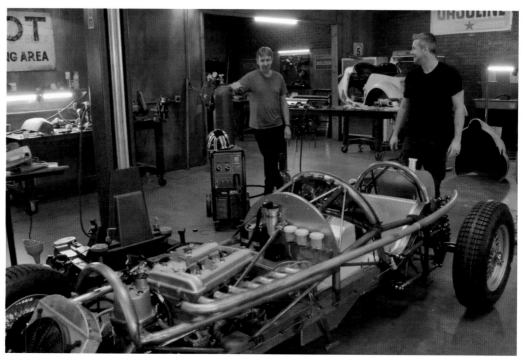

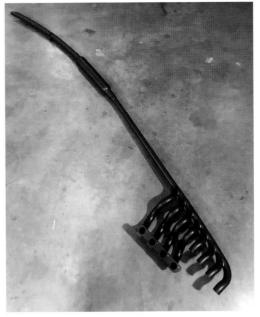

fully welded the silencer in position and ground down the welds.

To finish off, I shot-blasted the complete exhaust system, before having it treated with a satin-black ceramic coating. Ceramic coating is relatively new to the exhaust-coating business. It is applied as a powder and then baked, using a similar method to other powder-form paints. The ceramic coating is actually made up of a number of different materials, which create an incredibly thermally efficient barrier on the surface of the metal. This results in more of the exhaust heat being retained inside the pipes, which in turn means the gases travel faster, and the exhaust works more efficiently. Essentially, this ceramic coating makes the car go faster!

Finally, I added some heat-shield wrap to the front 10in of the manifold tubing, nearest the engine, to reduce heat in that area. I used jubilee clips to hold the wrap firmly in place.

When finally fitting the exhaust system, I made sure I used some new Alfa Spider exhaust gaskets to provide a perfect seal between the cylinder head and the manifold flange.

I soon realised that when I was sitting in the cockpit, the exhaust would be pretty close to my left arm, and it was going to get VERY hot. To keep things more comfortable, I rolled some 1.5mm aluminium sheet into a heat shield, welded on two ends, drilled with air holes, and then clamped the shield to the exhaust with jubilee clips.

That was my exhaust system completed – time to stand back and admire the hard work!

CHAPTER 24

Propshaft

Before going any further, I feel I should mention a little about terminology. I call the shaft that joins the output of the gearbox to the input of the differential the propshaft. Propshaft is short for propeller shaft, and the word comes from the shaft used on boats and aircraft to join the engine output to... the propeller! Other terms for this shaft could be driveshaft, cardan shaft, or tail shaft. I call the shafts that run from the differential output to the wheel hubs the driveshafts (though some people call those axle

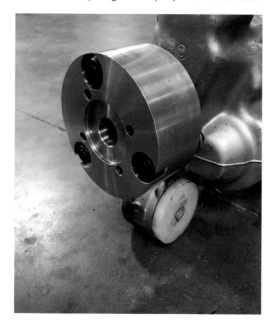

▶ **My version of the crack-prone rubber flex disc – a solid aluminium item machined to fit to the gearbox output flange and the propshaft.** *(Chris Hill)*

shafts). Right, now that's cleared up, let's get on with the propshaft.

The Alfa Romeo Spider donor provided a number of useful parts, and the fact the drivetrain is sourced from a single car is helpful. In an ideal world the propshaft would drop right in – however, this is not an ideal world. In my younger, more maverick days I would simply have 'cut and shut' the existing propshaft. For my special though, I decided not to cut corners, and to get a new propshaft made – it is not hugely expensive, and is a small price to pay for safety.

Starting at the front, I elected to remove the large rubber donut (flex disc) attached to the gearbox output flange that sits between the gearbox and original propshaft. This donut is like a low-tech constant velocity (CV) joint – it allows movement in a number of directions while still transferring power to the propshaft. The problem is that these rubber couplings can crack, so I got rid of it.

Instead, I machined a solid aluminium coupling that utilised a three-bolt pattern to match the gearbox output flange. I then drilled and tapped the face to accept the four-bolt pattern for a new universal joint (UJ), which I would fit to my bespoke propshaft. I bolted the coupling securely to the gearbox output flange.

I opted to get my bespoke propshaft made at a professional shop, which was fully kitted for the job.

There is a Universal Joint (UJ) at both ends of the shaft, which allows for movement, while still transferring rotational power. The middle of the shaft is in two sections, with a splined joint that allows the shaft to adjust in length slightly as the suspension rises and falls.

Propshafts like this are completely modular, and are welded together on a jig to hold all the components in the correct place.

Each universal joint is made up of a cross-shaped central piece called a spider, which has needle roller bearings for smooth movement, along with two yokes – one attached to the shaft, and the other to the gearbox output, or the differential input flange. The joints can articulate in all directions, while rotating and transferring power.

All the propshaft components were new, except for one – the rear UJ yoke that bolted to the differential flange. This particular part is hard to find, and the driveshaft shop that built mine recycled that part from the Alfa Spider.

▶ **A rotary welder produces a superb, neat and very strong joint when welding the propshaft.**
(John Lakey / Dunning and Fairbank)

Once all the parts were lined up, the operator slowly rotated the jig and welded the seams.

The shaft was then close to finished, but inevitably it wasn't perfectly balanced. Balancing a propshaft is essential for safety and for smooth driving. Even a minute discrepancy can cause a rattle and rumble when the car is being driven. Balancing the shaft for use on the car is done in two stages.

Stage one is tackled by using a micrometer, and fixing any imperfections in the shaft using heat from a gas torch. The shaft is spun by hand, and movement of the micrometer indicates any areas where the shaft is not running true. The gas torch can be used to heat the relevant part of the shaft, causing the metal to expand. The area is then 'quenched' using water, which rapidly shrinks the metal. This process can be used to fine-tune any imperfections until the shaft runs true.

Stage two involves spinning the shaft on a machine that replicates its use on the road (or track). The machine is able to determine where any small weights are required to balance the shaft. The machine can be used to indicate exactly where any weights should be fitted, and how heavy they should be. Any weights that need to be added are welded in place.

After one final balance recheck to make sure that it's spot on, a lick of paint finishes a propshaft that will last for years.

▼ **A final balance check on the shaft once it has been painted, and it is then ready to fit to the car.**
(Master Mechanic)

▼ **Finely balancing the propshaft using a purpose-built machine. Any balance weights required are welded in position.** *(Master Mechanic)*

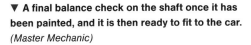

CHAPTER 25

Dashboard

▶ **I wanted to emulate the simple yet functional layout of the original Alfa 158 dash.** *(Ant)*

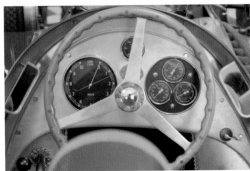

▼ **Paul's amazing sketch helped me lay out the dash, and get a feel for how it would look.** *(Paul Cameron)*

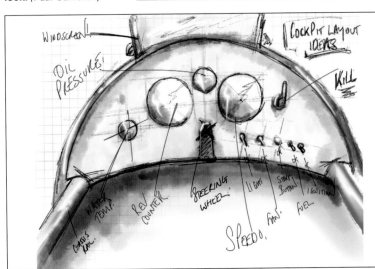

I intended to keep the dash clean and simple, just like the original Alfa 158 race car. This wasn't the type of car for any excess – less, as they say, is more.

As the original car was built purely for racing, the driver required only a specific amount of data. There was no speedometer, after all they didn't care how fast they were going, they just wanted to travel as fast as possible! The rev counter was large, and mounted in the driver's line of sight so he could maximise the rev range.

For my dash, I stayed with the racing theme and installed only the basics.

In keeping with the original, I made an aluminium panel from the same 2.5mm sheet I used for the cockpit. Using the same cardboard template I made to create the midway/cockpit chassis hoop, I cut and shaped the aluminium to suit. I drilled a hole for the steering column to pass through, then I drew around the inner edge of the chassis hoop and crossmember with a marker pen on the back of the panel. This showed me the area within which all the gauges had to fit to avoid interfering with the chassis tubes.

I raided the Spider donor vehicle for four gauges – the rev counter, speedo, oil pressure and water temperature gauges. I also saved the mechanical drive cables for the rev counter and speedo. One

◀ There was ample space behind my dash for the gauges and wiring, unlike today's cars where space is at a premium. *(Master Mechanic)*

of the advantages of the single donor was that the gauges were already calibrated for the engine and gearbox.

In addition to the instruments, I purchased four on/off toggle switches for the ignition, fuel pump, fan override and light circuits. I also purchased a red battery isolator ('kill') switch and a push-button starter switch.

Starting with the larger gauges, first I made horizontal and vertical pen marks on the dash to ensure correct spacing and an accurate layout, then

◀ This line marked on the rear of the panel (around the inner edge of the chassis hoop) is critical, as it shows the area in which all my dash-mounted equipment must fit. *(Master Mechanic)*

◀ All these components need to fit within the space marked on the dash plate. There are no rules, so the layout is up to me. *(Master Mechanic)*

▲ I simply drew around the gauges. If something didn't look right, I erased the line and moved it. *(Master Mechanic)*

▲ My workbench is steel, so I placed the dash on top of a piece of wood to drill into. Hole saws come in a variety of sizes, and aluminium sheet is relatively soft.

▶ I took time to make sure the gaps were equal and everything lined up as I wanted it. *(Master Mechanic)*

▶▶ I used an orbital sander to remove the shine from the aluminium to avoid glare. *(Master Mechanic)*

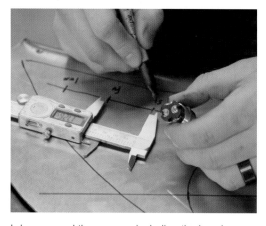

▼ I later identified each switch with an old-school labeller to give that authentic look. *(Master Mechanic)*

I drew around the gauges, including the bezels, one by one.

Using a hole saw in an electric drill, I drilled out each hole slightly under-size, checking the diameter

of the relevant gauge body before I drilled, to provide a nice snug fit (remember I marked around the front of each gauge – including the bezel – on the dash).

I then measured and marked out the positions of the switches at the bottom right of the dash. First, I drew a horizontal line at the right height for the centres of the switches, then I marked the switch positions at equal intervals using digital calipers. Next, I measured each one of the switches and its button, and drilled a suitable hole with a coned, step drill.

Next, to remove the shine of the aluminium (and remove the risk of glare on track), I cleaned off the pen marks and sanded the face of the dash panel down with an orbital sander. Finally, I gently filed each hole to give a nice, neat finish.

After carefully fitting each component to the dash and securing in place, I offered the complete dash assembly into position and drilled through the panel and through the first wall of the chassis tube using a 5mm drill bit. I then ran a 6mm tap through the chassis tube to take an M6 domed-head Allen bolt.

With the dash completed, and bolted in position, and the engine (and other electrical components) in place, the next step was to join them all together with a wiring loom.

CHAPTER 26

Electrics

Wiring... Just the very thought of electricity running through wires can put some people off. Don't be worried if that's you – I know a few good mechanics who can't wire a three-pin plug, yet can pull the engine out of a car in 45 minutes.

A modern, luxury car today is likely to have over 1,500 individual copper wires, totalling close to 1 mile in length, compared to a car of the 1940s containing maybe 50 wires, and about 150ft of wiring. A Toyota Prius today has about 30kg of copper wiring in total – I can't imagine tracking down an intermittent wiring fault on one of those!

The complexity of a wiring diagram for a typical modern car can look bewildering. However, the wiring diagram for my car looks relatively simple. It helps when you don't need any creature comforts, such as electric seats, in-car entertainment systems, climate control, or other fancy additions.

In the wiring diagram for the special, things are very much spread out for easy viewing. In the actual car there are two main clusters of wires – for the engine, and the dashboard. It's a bit like the London Tube map – if it was truly to scale, with all the stations the correct scale distance apart, you'd never be able to read it!

Wiring for lights is included on the diagram, as I'm planning to register my car for use on public roads at some point, and will need to fit lights at a later date. I decided to incorporate the wiring at this stage, to avoid having to adapt the loom later. If the car is never going to be used on the road, then the loom can be simplified.

Some basics

Electricity can be dangerous – even 12 volts – so it's necessary to observe a few precautions when working on electrical circuits. The battery might only be low voltage, but it contains a huge amount of electrical energy and supplies a high current (amps). A short circuit with a high current can cause very bad things to happen.

I wanted to keep everything logical and neat, so as I went along designing my loom, I checked and double-checked.

A basic electrical circuit is made up of three elements.

- A power source (a battery)
- A conductor (some wire)
- A load (a motor, or bulb, for instance)

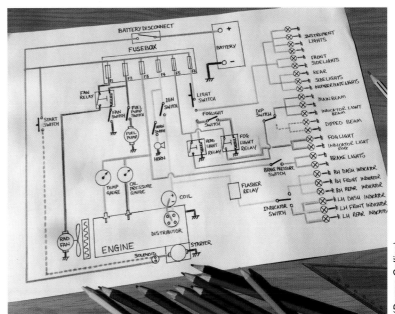

(Darren Collins)

▼ **The wiring in my car is clustered in two main areas – the engine and the dash. It's about as simple as car wiring gets.** *(Paul Cameron)*

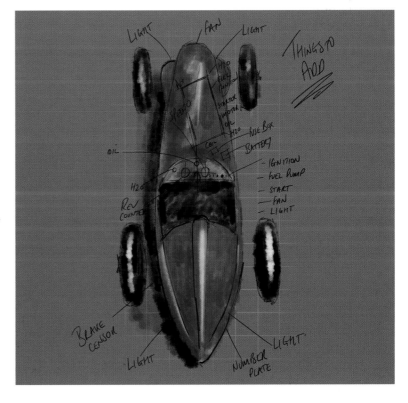

The battery has two terminals – positive and negative. Current flows from the positive terminal to the negative, with the load in between. This is a working circuit, and if left untouched will continue working until the battery charge is exhausted. Of course, circuits aren't always this simple – there may be switches, relays, capacitors, resistors etc.

In most automotive applications, including my car, the metal chassis, engine and gearbox form part of the electrical circuit. The metal parts of the car are used as part of the conducting path – the earth circuit. This simplifies things and means there are fewer wires to run.

On my car, I connected the black, negative terminal from the battery directly to an earthing point on the gearbox bellhousing. The engine was then separately earthed directly to the chassis using an earth-bonding strap, to ensure the circuit was complete, and not broken, or insulated, by the rubber engine mounts. This is normal practice on most cars.

This meant that I then only had to run a single wire to each load (component), as long as each load then had a wire connecting to the chassis (its own earth point), which completed the circuit to allow the electrons to flow.

Selecting wiring

I used standard single-core copper automotive wire, but it's most important to select the correct size (cross-sectional area) of wire and the correct fuse (rating in amps) for each circuit. If the wire is too small, the electrical current (amps) has difficulty travelling down the wire, causing a kind of 'electrical

pressure' to build up. This is called resistance and results in the wire getting hot – too small a wire with too large a current can cause an electrical fire and that's really not good. Equally, if your wire is too big, it results in unnecessary extra weight and cost.

Some of my parts, like the cooling fan motor, came with manufacturer's data which told me the working current they required, but because I was building a special, there were certain electrical circuits specific to my car that I wanted to wire in which did not come with instructions. So how did I know what size of wiring was safe to use for these?

In an electrical circuit, the Power Law states that power (W, or watts) equals current (I, measured in amps) multiplied by voltage (V or volts):

$$W = I \times V$$

So, to work out the current in a circuit, we can divide the power by the voltage:

$$I = W \div V$$

As an example of how to work out the wiring size needed, let's look at the lighting wiring circuit. Most car light bulbs have a current rating, in watts, engraved around the bulb's metal ferrule, and often it's shown in the car's handbook too.

For my car, I wanted wiring for front and rear sidelights. Each of the bulbs was a 5-watt bulb, and on the same circuit I had the number plate light (5-watt bulb) and the dashboard gauge illumination lights, which were 3-watt. So, I needed to add up the total number of watts for all the loads (bulbs) on the circuit, and put the total into the Power Law equation.

Here goes:

    2 x front side lights (2 x 5W) = 10W
    2 x rear side lights (2 x 5W) = 10W
    1 x number plate light (1 x 5W) = 5W
    6 x gauge lights (6 x 3W) = 18W
    Total electrical power in circuit =
        10 + 10 + 5 + 18 = 43W

To work out the current (I, in amps) I simply divided the total number of watts (power) by the voltage (V), which for most cars is 12V:

$$I = 43 \div 12 = 3.58A$$

So the current rating for my sidelight circuit needed to be 3.58A

With this information, using a wire chart, I could now select the correct cross-sectional area of wire to feed this circuit, and I could also select the correct fuse rating. The wiring chart shows the required cross-sectional area of the wire, in mm², depending on how long the wire needs to be.

▶ **Wiring charts like this are very useful. When you have worked out how much current is likely to flow, you can select the correct gauge of wire.** *(Darren Collins)*

12 Volt								
acceptable cable size (mm²)								
	Cable Length (metres)							
Amps	1	2	5	10	15	20	25	30
0.5	0.4	0.4	0.4	0.4	1.84	1.84	1.84	1.84
1	0.4	0.4	0.4	1.84	1.84	1.84	1.84	1.84
1.5	1.84	1.84	1.84	1.84	1.84	1.84	2.9	2.9
2	1.84	1.84	1.84	1.84	1.84	2.9	4.6	4.6
3	1.84	1.84	1.84	1.84	2.9	4.6	4.6	7.9
4	1.84	1.84	1.84	2.9	4.6	7.9	7.9	7.9
5	1.84	1.84	1.84	4.6	4.6	7.9	7.9	13.6
7.5	1.84	1.84	2.9	4.6	7.9	13.6	13.6	25.7
10	1.84	1.84	4.6	7.9	13.6	13.6	13.6	25.7
15	1.84	1.84	4.6	13.6	25.7	25.7	25.7	32.2
20	2.9	2.9	7.9	13.6	25.7	25.7	32.2	49.2
25	4.6	4.6	7.9	25.7	25.7	32.2	49.2	49.2
30	4.6	4.6	13.6	25.7	32.2	49.2	49.2	
40	7.9	7.9	13.6	25.7	49.2	49.2		
60	13.6	13.6	25.7	49.2				
80	25.7	25.7	25.7	49.2				
100	32.2	32.2	32.2					
125	49.2	49.2	49.2					

Most automotive suppliers sell wire in sizes such as 0.5mm^2, 1.0mm^2, 2.5mm^2, 4mm^2, 6mm^2 etc, so it's best practice to round up to the next size up if the chart results fall between wire sizes. I worked out that the current in my lighting circuit would be 3.58A and the distance between the fusebox (the starting point for the circuit) and the front and rear lights was around 2.5m. Looking down the 'Amps' column in the chart took me to 4A (rounded up from 3.58A) and then across to the 5m column (again rounded up from 2.5m). Bingo! I needed a minimum wire cross-sectional area of 1.84mm^2, so probably 2.0mm rounded up.

I bought various different colours of wire to make it easier to follow the wiring runs through the car (from end to end) in case I had to diagnose any problems later on. There is nothing to say that you can't use the same colour wiring throughout the car, but unless everything is very clearly labelled, it would be a big problem working on it at a later date! It's also worth bearing in mind that it's common practice in most cars to have the negative ('ground' or 'earth') wiring black throughout, and the main positive running from the battery to the master switch, and from the master switch to the fusebox, red. There are no hard and fast rules for the rest.

Fuses and Relays

With the wire sizes carefully selected for each circuit, next I needed to think about fuses and relays. Remember, I mentioned previously that if the current has trouble travelling down a wire that's too small, the wire can get hot, and even cause a fire? Well, the same thing can happen if there is an unexpected increase in the electrical load on the circuit. Typically, this can happen in a dead-short situation, for instance, if a wire's insulation rubs through and touches the chassis, or another earthing point. In that case, all the charge the battery holds wants to try and travel down that cable (the easiest route) instantaneously. Definitely not good! That's why fuses are required – they protect each circuit if an overload situation occurs.

A fuse or fusible link is a small section of wire made from tin, lead or zinc that will comfortably carry current up to a predetermined level. If this set level is exceeded, the fuse will heat up, melt ('blow') and break the electrical circuit, saving the wiring and components in that circuit, and lowering the risk of more serious damage.

The fusebox is a central holder for the car's fuses. The battery positive feed is connected to the fusebox via the master switch, and the main feed is then divided to feed each electrical circuit via a fuse.

Like the wiring, it's very important that the fuses are rated correctly for each circuit fed from the fusebox. Each fuse needs to protect the relevant load and the circuit wiring, so as a rule, a fuse

▲ **Off-the-shelf fuse/ relay boxes like this are easy to source, and some come loaded with fuses.** *(Ant)*

should have a **higher** rating than the total current rating of the circuit (remember the light-bulb calculation?), but a **lower** rating than the wire rating, so that the fuse blows before the wire melts.

As with the wiring, fuses come in set ratings and should always be selected to protect the smallest cable in the circuit they supply.

In my lighting circuit, I worked out previously that the bulbs drew 3.58A, so the fuse selected for that circuit needed to be just about this rating to stop unnecessary blowing. Car fuses are generally available in the following sizes: 2, 3, 4, 5, 7.5, 10, 15, 20 and 30A. The ideal fuse for the sidelight circuit would be 4A, and according to the wiring chart, the 1.84mm^2 wiring is good for 5A, so with a 4A fuse, I had the circuit protection (the fuse would fail before the wire).

With the fuses and wiring selected, the final electrical safety device I needed to consider was a relay. Relays allow a low-current circuit to switch a higher-current circuit. They are useful when the circuit you want to switch has a greater operating current than the switch you've chosen can handle. A relay has two separate, isolated electrical circuits. The dashboard switch is used to turn on and off a low-current coil inside the relay that acts as an electromagnet. This electromagnet then pulls a metal contact to operate a much bigger switch inside the relay, which turns on or off a high-current load. The only high-current circuit for which

▶ **The battery has to be securely fixed. It must be rigidly mounted using a clamp.** *(Chris Hill)*

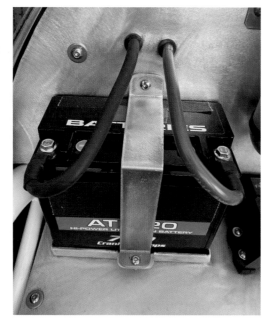

▼ **The finished wiring loom, neatly trimmed and wrapped. Car wiring looms don't get much simpler than this.** *(Ant)*

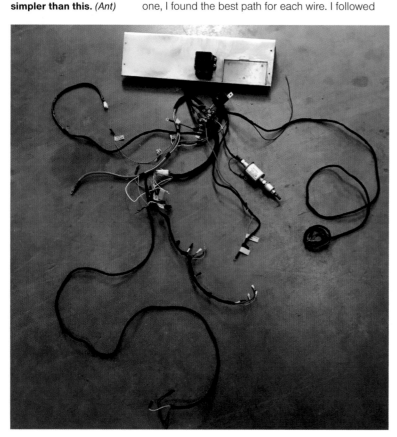

I needed a relay on my car was the radiator fan circuit. As with the other electrical components, relays have current ratings, so it's important to select the correct type based on the requirements of the circuit it feeds.

Making the wiring loom

There is no secret to routing the wires. One by one, I found the best path for each wire. I followed the wiring diagram I had created as a guide, and crossed off each wire once I had installed it. I left a good few inches at each end to allow for any movement or re-routing, as I could trim the wires to length later. I labelled each wire, using masking tape, so that I would know where to reconnect it later.

It was necessary to drill holes in the engine compartment bulkhead for a few wires, and at this point I had to bear in mind that I would be fitting rubber grommets to protect the wiring insulation, so the hole in the bulkhead had to match a suitable grommet size. I had a number of different-sized grommets available so I could pick accordingly.

As more and more wires were added, the bundles grew. I temporarily cable-tied wires together in groups to keep things manageable. It always looks neater if you group certain wires together, until they branch off to their own load.

Once I had worked through the whole wiring diagram, and all the wires were in place, it was time to strip them all out again. I used cable-ties to keep groups of wires together at the points where certain wires separated and headed in different directions, to make things easier for the next step.

I took some cloth insulation wrapping (PVC electrical insulation tape does the job too) and started to wrap the tape around the bundles of wires. I chose cloth tape, as it is a little more flexible than PVC and was more in keeping with the period look of my car. This is probably one of the most satisfying jobs to do well (and the most annoying when it isn't done well). It's worth taking some time, and often best to ask a helper to hold the wires while you wrap the tape carefully. When I arrived at each of the cable-ties, I carefully snipped it off and replaced it with the wrapping tape. At this point, the bundle of wires officially became a 'loom'.

When the loom was completed, I fitted it to the car, and secured in place with clips to keep it neat.

Wiring connections

There are a couple of ways to connect the end of the wire to the component it's powering. Crimped terminals are probably the most popular. If the correct size of terminal is selected for the wire gauge, and a good-quality crimping tool is used, these are just fine. It's important to ensure that each connector is very firmly crimped, and it's an idea to test each connection by pulling on it lightly – better to find out now rather than later.

Some people insist on soldering all the terminals, and that's also fine. In this case, a section of heat shrink should be slipped on to the wire before soldering, then pulled over the soldered joint before shrinking with a heat gun to keep it insulated.

With the wiring completed, the car is beginning to come to life!

CHAPTER 27

Seat and harness

Seat

Cars from the 1930s had incredibly basic interiors when compared to today's modern designs. What strikes me most is the incredible advances made in comfort and safety. Modern bucket seats are built to save the driver's life, whereas the seat in the original Alfa 158 was merely built to seat the driver in such a way that he could control the car at speed.

The original 158 that I sat in had a single bucket seat trimmed in beige corduroy! Yes corduroy! It looked like a geography teacher's jacket!

At Goodwood Members' Meeting in 2018, there was a beautiful single-seater, boat-tailed Bentley special. It didn't have a seat per se; instead, the cockpit space was trimmed in rich brown leather. She looked sensational, and in the ethos of special building, I am taking inspiration from that Bentley, and not from the original Alfa.

The first stage in the seat-making process was to select the materials. I started by selecting a worn, vintage brown leather, and complemented it with ivory thread to match the chassis colour.

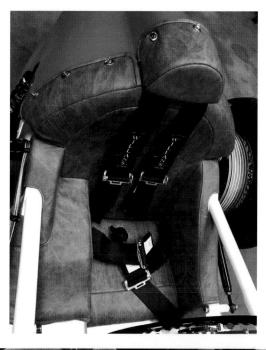

▼ The original 158's seat looks more like a school teacher's jacket than a racing seat. I love seeing that lightweight chassis. *(Ant)*

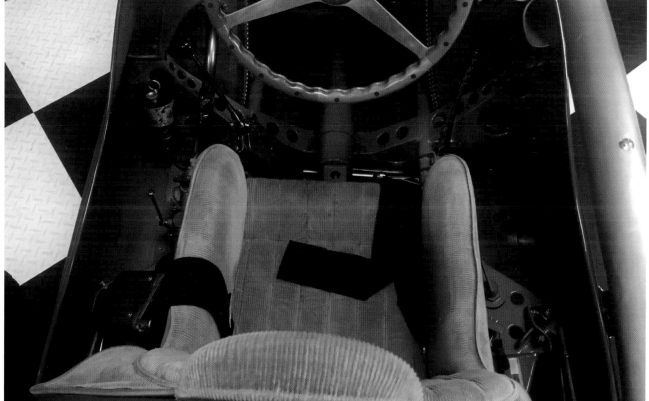

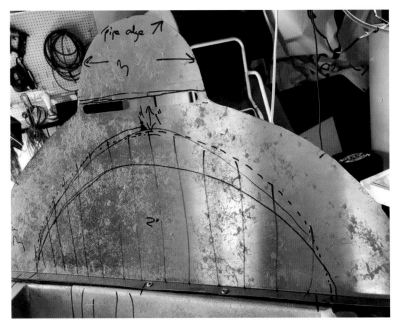

▲ **I simply drew on the aluminium to plan the interior upholstery, with lots of getting in and out of the car to make small alterations.** *(Ant)*

The fluted areas of leather were glued to some thin 5mm foam backing, then chalk lines were drawn 2in apart. Ivory stitching was then sewn along those lines. The procedure was repeated for each panel.

Bit by bit, panel by panel the seat came together. The fluted centre sections and curved fluted side panels were designed to give a quality, snug feel to the seat, and of course to support me when driving the car.

The leather trim was designed to wrap around the chassis bars to attach to hidden fixing points along the outside of the chassis rail. To check the fit of the seat components, I frequently trial-fitted the parts to the chassis/body.

To secure the leather trim, I pulled the edges over the chassis, and used turnbuckles to secure the edges. To help to secure and spread the leather, we also sprayed adhesive on the back of the leather.

To fix the leather to the tail section, I used turnbuckle fasteners riveted to the edge of the tail section, pushed through eyelets fixed to the leather. This provided a nice detail on the finished car, and also meant that the trim was easily detachable from the tail section – essential if I needed to remove the tail.

The finished seat was incredibly comfortable and looked fantastic against the cream chassis, aluminium cockpit and red body.

Harness

The idea of a harness in a car like this is pretty foreign. They simply didn't have any in period! And actually, that made the car safer, because as you started to roll you stood a chance of reducing

I then called in my pal Revo to trim the seat for me. I set about drawing the details I wanted directly onto the aluminium tub with a marker pen. I wanted the curve of the backrest to follow the line of the chassis. I also decided that I wanted the flutes of the seat centre section spaced 2in apart, with a 3in plain front-edging section under my knees.

The fact that I made the aluminium tub removable was a real asset at this stage, and meant that the job could be tackled on the bench. Revo set about adding different thicknesses of foam to the structure, based on my needs – 2in for the backrest, 4in for the headrest and ½in for the side panels.

▶ **The different-thickness foam shapes are glued to the metal panels.** *(Revo)*

▶▶ **Working with soft, brown leather is a big change from steel and aluminium. It's like adding the curtains after building a house from scratch.** *(Revo)*

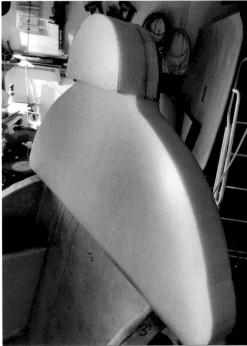

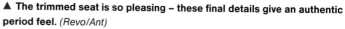

▲ **The trimmed seat is so pleasing – these final details give an authentic period feel.** *(Revo/Ant)*

injury by diving out – yes, jumping from the car! Having a harness fitted meant you were likely to get trapped in the seat, and with your head above the body line, and no roll bar, you can imagine it would get very tricky very quickly.

I also don't have a roll bar that sits above my head, but I have created a car with far more stable handling characteristics than the original 158, so I have made an educated risk assessment that it's far more likely that I would crash with my wheels planted on the ground (head-on, rear-ender, side impact) than that the car would roll. So I opted for a harness to protect me in the event of an impact.

The type of harness fitted will depend on use of the car, and the local regulations. Always check with the regulations that apply for track use, which can be quite different from what is required for road use.

I fitted a four-point quick-release harness. The two upper straps attach to two simple brackets welded to the space-frame bar behind my shoulders and the straps pass neatly between the seat back and the headrest. The two lower straps were attached to brackets in the lower hip area, welded to the chassis.

◀ **With these turnbuckles, I can still remove the whole seat in one section if required.** *(Ant)*

◀◀ **The finished seat has a real quality feel. It's important to keep up the professional look, as these are the parts that people will ultimately see first.** *(Ant)*

CHAPTER 28

Preparation and painting

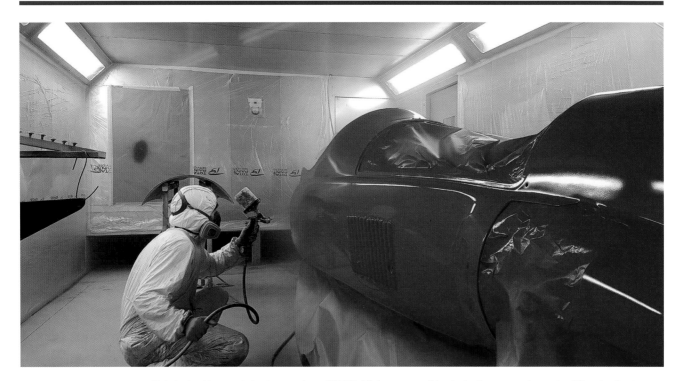

This is the bit people look at and say WOW! All the hours put into engineering the rest of the car will be wasted if the final finish is poor! The paintwork is essential, and the car will be measured by it! Remember, 90 per cent of the work of painting the car is in the preparation.

Preparation

So, did I need to rectify any small imperfections? If yes, now was the time! I crawled over the bodyshell in forensic detail to seek out any pinholes, inaccurate body lines, deep scratches or ripples. If I found any, they would need to be put right, so if necessary I would need to get out the body filler, rectify the problem, check it, check it and then check it again!

The body was made with a gel-coat layer for a reason, and it was extremely important not to break through this during the preparation for painting – I really didn't want to expose the glass fibres underneath. I kept the gel-coat layer sealed, as it would prevent problems when it came to eventually priming and painting.

I was incredibly critical of my bodyshell, which, in fairness, had already had a short, but tough, life. Not only had it travelled the 6,000-odd miles from my UK workshop, where the buck, mould

and bodyshell were made, to the *Wheeler Dealers* workshop in California, it had also been used as the 'jig' to fabricate the space frame. It had been drilled, cut, bashed and banged, and the odd scratch was inevitable, but that's the beauty of composites – they can take a beating!

I set about ensuring the shape was flawless, using some thin layers of filler. The areas where I had joined the cooling louvres to the smooth shell needed some attention to ensure the join lines were invisible. Also, the centre of the car – where the mould halves joined – inevitably had a thin join line. That, too, needed some gentle attention.

It was like being back at the 'buck' stage in Chapter 13 – applying thin layers of filler and knocking back over and over until I was happy.

The preparation work at this stage is most important – extra time spent now can save hours later, as ultimately the car will be judged on its final finish. This is not the time to cut corners.

The body, due to the nature of its construction, is reasonably flexible, and left unsupported over time could sag or distort, so it is critical to make sure that any adjustments to the panels, shut lines and joints are done with the body supported as it will be when fitted to the car. For me, the answer was simple –

leave it on the chassis. My chassis was going to be blasted and powder coated later, so it didn't matter if it got a little messy during the body preparation, although I did decide to mask up the main visible areas to make light work of any cleaning up later.

So, with the body supported on the chassis, I could check the fit of the bonnet and the overall shape of the bodyshell, to ensure that the curvature and shut lines looked parallel, symmetrical, straight and matched. When you look along a car or body panel, you can often see instantly if there is a defect, or a problem with a gap, dent or ripple. Starting with these imperfections first, I keyed the surface all over and applied a thin skim of filler in the areas needing attention, sanding them into shape. Due to the way gel coat and glassfibre is applied to a mould, it was inevitable – despite best efforts – that I would have the odd tiny air bubble trapped somewhere. Some of these were easy to see with the naked eye, some I could feel with my fingertips, and some I discovered when keying the surface for primer. The important thing was to fill or 'stop' these bubbles, as they would show as pin holes or blisters in the final paint finish.

With the body held rigid on the chassis, and the bonnet fitted, I could see the panel gaps clearly. Looking along the gap from one end will accentuate any curve, wave or distortion in the gap between the bonnet and body. I used a thin length of masking tape held taut while I applied it along the edge of the shut line, to show where any work was needed.

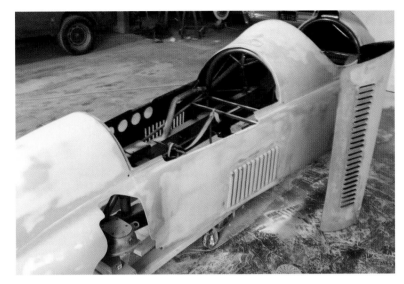

Now I could decide whether to take a little off the bonnet edge where the gap was tight, or add filler to the body where it was wide. In reality, it took a little of both to get things just right. That's the beauty of working with glassfibre – unlike steel or aluminium, it's easier to add and take material away, and if you get it wrong first time, it's possible to roll the clock back and start again.

When all rectification work was complete, the next job was to give the body a 'key' to enable the primer to stick to it. Any shiny gel-coat surface left unkeyed will cause the primer and paint to separate later, as it will not adhere correctly.

▲ **After another very light skim of filler and a sanding back, the bodyshell is ready for the paint booth.**
(Sean Winograd)

◀ **I used yellow masking tape to ensure each shut line and edge was perfectly straight.**
(Sean Winograd)

▶ **Primer coat. This is when you really get to see whether the body has any imperfections. This is the stage where the various curves of the body blend into one unified shape.**
(Sean Winograd)

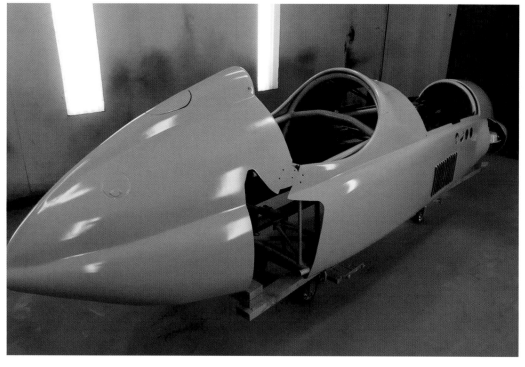

To carry out the flatting, I used P400-grade discs on a DA (Dual Action) sander with an interface pad. As there wasn't a truly flat surface on the curvaceous body, the interface pad – a flexible 'sponge' – allowed the abrasive disc to follow the body contours. As with all sanding jobs, a dust extractor is required for this, along with a quality dust mask or respirator. Sanding should always be carried out in a well-ventilated area.

Once prepared, I wiped over the complete body with a cloth, then used a vacuum cleaner and tack rag (a special wiping cloth made from tacky material to remove dust and lint). I then had one last look over the body for any imperfections, and rectified them before priming.

Painting a car is an exciting and very satisfying moment, as all the hard preparation work comes together and the body changes from a collection of work areas to a unified form.

Primer

It was now time to prepare for two to three coats of high-build primer. High build is exactly what is required on a gel coat and glassfibre body, to cover evenly and provide enough material to cover, seal and allow final flatting before the colour coat. So, I was ready to get in the spray booth! The procedure here was the same process described in Chapter 13 (Building the body buck), building up layers of filler and sanding.

The first step was to get suited and booted and degrease the body using a solvent-type panel wipe, then I made use of the time while any solvents evaporated to mix the high-build primer following the manufacturer's instructions.

I gave the body a final wipe over with a tack rag to remove any dust particles, and then applied the first coat of primer. It's important to allow time between coats, and not rush, in order to let the primer dry ('flash off'). I gave it 10–15 minutes between each of three coats.

It's important not to 'bake' a GRP shell in order to dry it, as this can damage the glassfibre. It MUST be air-dried.

I used trestles to support the bonnet, so that it could be painted off the car at a comfortable working height. This also allowed for

▼ **The bonnet is handled in the same way, placing it on trestles to work at a decent height.**
(Sean Winograd)

the bonnet shuts to be accessible on the main body for painting.

It was essential to make sure that the primer coat was dry before attempting to prepare the primer for colour – the hardener creates a chemical reaction with the primer, requiring 24 hours to cure depending on atmospheric temperatures. A good way to test if it's ready is to gently dig a fingernail into an unseen area. If it leaves a mark, walk away – do not rush it!

When the primer had cured and was hard enough, I 'wet flatted', starting with P800-grade wet-and-dry paper, working through to P1200, using a rubber block by hand. P1200 grade gives a surface smooth enough to apply the colour directly over the top.

A neat trick at this stage is to wet the surface with a cloth and take a good look over the body. A thin layer of water acts just like a coat of very glossy paint, so you'll see any ripples or imperfections not previously visible as they shine out on the wet surface. It's much easier for these to be dealt with at this stage before colour is applied.

Colour

When I was happy that the preparation was as good as it possibly could be, it was time for the real fun part – colour!

As with the preparation for the primer, it was even more important now than before to ensure the body was free from dust and grease, so I went over it several times again with degreaser and a tack cloth.

I wanted my car to really stand out, and I decided to add a yellow and blue 'moustache', not only to accentuate the beautiful body lines and complement the main-body red, but also in homage to my hero, Fangio, who had the blue and yellow stripes on the nose of his 158 to reflect the national flag of his home country, Argentina.

To apply colour combinations effectively, it's important to apply them in order, starting with the lightest. Since the main colour was going to be red, I had to be extra cautious not to paint over this colour, as it is notorious for 'bleeding', or showing through other lighter colours. For this reason the yellow was painted first, then the blue and finally the red.

As with getting the shutlines straight earlier in the preparation process, it was critical to get any stripes, or in my case the 'moustache', level and straight on the car. Unlike a vinyl decal, which can be cut out symmetrically, I was relying on lines produced by applying masking tape to the body.

After a lot of measuring, I marked out the rear edge of the 'moustache' with a fine-lining tape (blue tape in the photographs), then used three lines of 1in masking tape (yellow in the photographs) side by side, following the fine-lining tape to ensure that my moustache lines were exactly 3in wide and parallel over the curvature of the body. Once I was happy with the masking, the rest of the body was protected with masking paper (held in place with

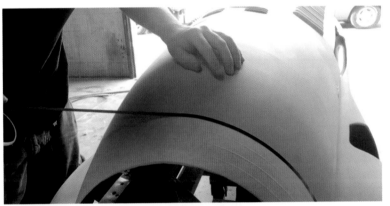

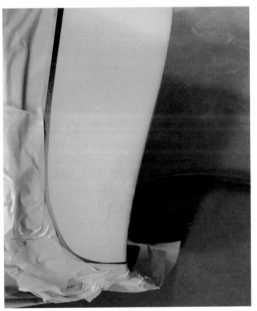

▲ The 'moustache' – yellow and blue to match Fangio's car. *(Sean Winograd)*

◀◀ The finished 'moustache'. In the *Master Mechanic* show you'll see Chip Foose argue against this. I went for it anyway! *(Sean Winograd)*

◀ Properly preparing for painting is more of a job than actually laying on the colour. *(Sean Winograd)*

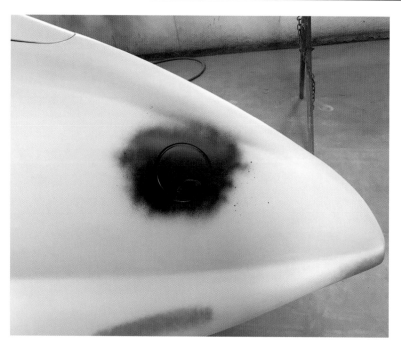

▲ **Some detail lines on the body were sprayed black first and then masked off. This provided nice sharp detail when the main colour was sprayed.**
(Sean Winograd)

▼ **The completed bodyshell with colour basecoat applied.**
(Sean Winograd)

so that it could be revealed after the main colour was sprayed.

The paint I'd chosen was a two-stage process – a colour basecoat followed by a clear lacquer topcoat. There are single-coat 'solid' paint products available, but with my yellow and blue racing moustache already painted on the car, a final lacquer coat would seal them all in, creating a beautiful smooth finish across the stripes onto the main colour.

The colour basecoat has a lower viscosity than the high-build primer, which means it will run if applied too slowly or heavily. So, with this in mind, it was necessary to keep my spray passes light and quick during the first coat. I was not trying to create a shine, as the basecoat dries dull anyway. I built the layers up, finishing with full coats, and allowing adequate flash-off time between coats.

I used the flash-off time to have a good look over the body while I was waiting, making sure there were no dry edges or missed areas before moving on to the clear coat.

Clear coat

After allowing the basecoat to dry thoroughly, I unmasked the moustache and fine-line tape details.

After one final check-over for cleanliness, I was ready for the clear coat. The clear coat application involves spraying a 75 per cent coverage coat, then waiting for it to flash off before applying a full coat, waiting for that to flash off, then applying a final full coating.

Because the coating was clear, I used the spray-booth lights as a guide to see a reflection on the surface, bending down and looking from all angles

more masking tape), and the yellow basecoat paint was applied. I used yellow paint to match Ferrari Galio Fly, paint code 20-Y-191.

I allowed the yellow to dry thoroughly, before the same masking process was used for the blue. Once dry, the two colours were covered to protect them before moving on to the fine-detail bits.

I wanted to highlight the fine details of the body moulding, so they looked sharp after applying the main body colour. This was done by simply spraying aerosol satin black over the details. After allowing the black spray to dry overnight, I used fine-line tape in the detail areas to effectively mask the black off

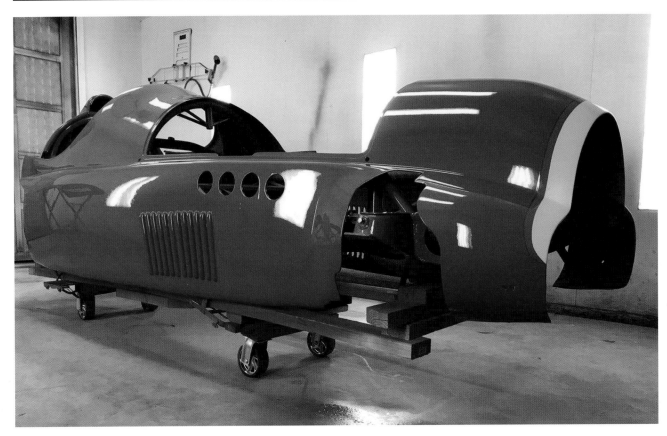

to make sure the coverage was even during each application. Once I was happy, it was important that I left the paint to dry until rock hard. Only then could a final flat and polish be carried out.

Final flat and polish

A modern spray gun is designed to replicate a modern car paint finish using modern products, so it will leave an orange-peel effect regardless of how much you try to eliminate this. If an attempt is made to thin the colour or clear coat down to make it flow better, this could result in longer drying times, heavy runs, sags and possible future blistering of trapped solvent, so not desirable.

I wanted a period, high-quality glass-like finish on my car, so to achieve this, a flat and polish was needed. This means basically sanding smooth the surface of the paint to remove the orange-peel surface, followed by polishing it back to a high gloss.

To do this, I needed to protect all the edges, corners and detail areas by applying masking tape to stop me accidentally sanding through the paint, which is very easy to do on an edge.

I carefully wet-flatted, using a fine P2000 wet-and-dry paper, drying off in between to monitor the progress of orange-peel removal. Working on a section at a time until all the tiny peel spots were only just removed, I worked around the whole body. At the same time, I carefully flatted out a few runs and small dust particles using a rubber block.

Finally, I removed the tape from the masked edges, and carefully polished the clear coat to a very high gloss, using a series of cutting compounds, then mop-finishing with a protective glaze polish.

Finally, I stood back and admired the finished result of all the hard work!

▲ **The final layers of clear coat transform the finish.** *(Sean Winograd)*

◄ **The finished bonnet to match. After a final flat and polish the finish is amazing.**
(Sean Winograd)

CHAPTER 29

Brightwork

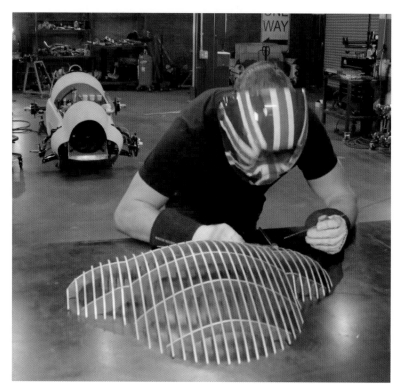

I love the fact that the original 158 has very little bling. The additions to the car are minimal for a reason – it's a race car, weight is an issue and, of course, the aim is to go fast, not look pretty (albeit the car is achingly beautiful). However, the 'shiny' parts of the car are still important, and also serve a purpose.

Aeroscreen

As you may remember, I fitted the aeroscreen when I was building the space frame, in Chapter 16. Now, I wanted to add a finishing touch in the form of a leather 'gasket' to sit between screen and body, to give that area a little extra feature, and also to swallow any minor imperfection in the fit between the two.

I removed the four aeroscreen securing screws and lifted the screen from the body.

I then made a cardboard template to match the shape of the aeroscreen mounting flange, but with extra clearance all round (I wanted the 'gasket' to be visible under the aeroscreen). I wrapped the card in leather left over from the seat covering. To add that final touch, I added a stitching detail to the leading edge and sides.

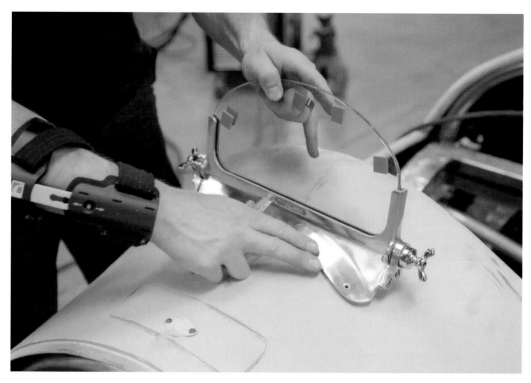

▶ **I removed the aeroscreen from the body, ready to fit the leather 'gasket'.**
(Master Mechanic)

Finally, I drilled holes through the leather to match the positions of the screen-mounting holes, placed the 'gasket' on the body and refitted the aeroscreen.

Bonnet handles

The original 158 has a pair of 'T'-shaped bonnet handles fitted to assist the removal of the bonnet.

Although I toyed with keeping the car authentic, having walked round a number of kitchen hardware shops looking for options, I managed to find numerous 'T'-handles that would bolt straight on, but my eye was immediately drawn to something a little different.

I saw some shaped and polished 'O'-handles that just spoke to me. I loved the way they looked and felt, and fitting them to the bonnet with two small holes and small bolts was simple.

Bonnet springs

Securing the bonnet could not have been simpler! A set of four bonnet springs and hooks was simple to source. It's important that the springs and clips are fitted at an angle to pull the bonnet into the corners of the engine-bay opening and stretch the load outwards.

I marked the positions for the spring-securing bolts on the body, and placed some 4mm mild-steel triangles behind them. These fitted into the recesses between the space-frame hoops and the tube side members. I clamped the triangles to the bodyshell and tacked them in place on the tubes. I then drilled and tapped an 8mm hole at each of the marks, through the bodyshell and into the metal triangles.

▲ **Leather gasket fitted. It was made from the same material as the seat, and adds to that period feel.** *(Ant)*

◀ **I really fell in love with the look of these bonnet-lift handles!** *(Ant)*

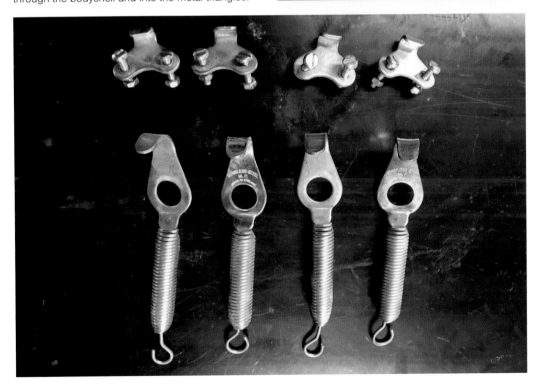

◀ **The bonnet springs and hooks.** *(Ant)*

▶ I didn't want the bonnet springs rubbing on the paintwork, so I machined these little spacers. *(Ant)*

▶ Lifting the bonnet on and off is pretty much a two-person job. Its big. *(Ant)*

▶ The spacers work really well to stop the spring coming in contact with the paint, and they look the part. *(Ant)*

I also machined a small spacer from aluminium on a lathe, ¾in diameter and ¾in tall, with a taper and an 8mm hole. This lifts the spring away from the paintwork to protect it, and looks great too.

I used nuts and bolts to fix the spring hooks to the bonnet, but rivets would have done just as well.

Grille

There was only one way to source an authentic-looking 1930s Alfa 158 grille and that was to make it.

Back in the 1930s, each grille would have been hand-made in aluminium. Bearing in mind the 158

158 DESIGN AND THAT BEAUTIFUL GRILLE

The Alfetta had two, perhaps three if we are being pedantic, body shapes. It was originally designed with a three-piece grille and exposed front suspension. For a short period that basic design was revised into a single-piece grille.

However, at the Coppa Ciano in 1939, the 158s appeared with totally new, and much more aerodynamic bodywork, which closed in the front suspension. Appropriately, Farina won, taking his first victory aboard a 158, but more importantly the car was now in its final body form. The one-piece grille, covered front suspension, small headrest, integral mirrors and beautifully elegant tail, would remain fundamentally unchanged until the car was retired at the end of 1951. For me, the later shape was a massive improvement – it's that later version I fell in love with and wanted to build.

(Centro Documentazione Alfa Romeo – Arese)

was a race car, and contact between cars was pretty frequent, I wonder how many grilles were actually made for the cars Alfa Romeo built and rebuilt over the life of the car?

Having the CAD data for my car prepared by Paul meant that it was a relatively straightforward task for me to reproduce the grille.

Modern-day laser cutting is really simple and accessible. I forwarded the CAD file to the laser cutter, and the grille components were cut in minutes from a sheet of 3mm aluminium. Note that each grille slat is designed with a slot so that it all interlocks together.

Having assembled the grille, and ensured that the components were straight and secure, I TIG-welded each joint, front and rear, with a small tack-weld. These tacks are incredibly strong and make the grille a substantial structure.

Offering the grille up to the nose section of the bodyshell revealed a lucky coincidence – the chassis horns on the MG TD ladder chassis were

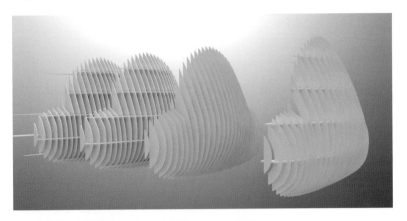

▲ A CAD model of all the grille pieces makes it a breeze to get the parts CNC cut. In Fangio's day it would have been very labour intensive to make this part of a car. *(Paul Cameron)*

▼ **I laid out the components, and then simply slotted them together to build a self-supporting structure.** *(Master Mechanic)*

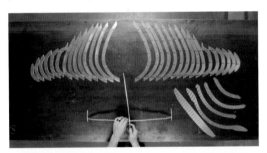

◀ The grille was designed to provide a flat rear, allowing it to sit on a workbench. Easy! *(Master Mechanic)*

◀◀ The MG TD chassis horns provided perfect mounting points for the grille. *(Chris Hill)*

◀ I welded tabs onto the grille, then drilled holes to align with the holes in the chassis horns. *(Chris Hill)*

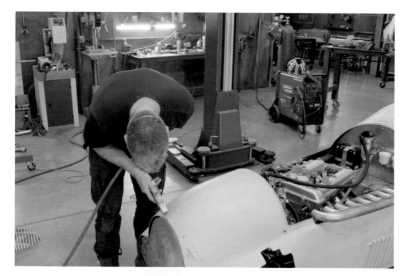

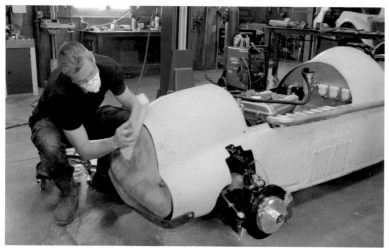

perfectly placed for mounting. I marked the chassis-horn location on the grille, then made two small tabs, drilled with 6mm holes, and welded them in position on the grille, so that they lined up with the holes in the chassis horns.

I then TIG-welded a 6mm nut to the inside of each chassis horn, behind the existing holes. I could now bolt the grille to the front of the car using two 6mm Allen bolts.

Straighten the nose

When I made the mould for the body, I didn't have the grille, so I made the fibreglass nose a touch too long, knowing I would trim it down at a later date. I made a template of the grille shape from 18mm plywood, which slotted nicely into the hole in the bodyshell. I positioned the plywood where I wanted my finished grille to sit, then I marked around the wood to indicate where to trim the nose. It didn't take long with my little air-powered saw. A quick sand with a block and some 100-grit abrasive paper, and the nose was trimmed ready to accept the grille.

I then added a flat piece of 2mm aluminium with a 1in hole drilled in the centre to provide a faux crank-handle hole, to mimic the original car.

I was thrilled with the final result – the grille is a vital part of the character of the car. I gave it a quick clean in my soda blast cabinet to remove the shine of the aluminium and to dull it down a bit.

With the yellow and blue 'moustache' (which helped the Alfa team distinguish which car was which – yellow and blue for Fangio, my favourite driver) it really pops!

▲▲ **The front of the car needed a little trim. Easy with an air saw and a plywood template.** (*Master Mechanic*)

▲ **I then used a sanding block and 100-grit abrasive paper.** (*Master Mechanic*)

▶ **The finished grille looking splendid, ready to attach. Note the fake starting-handle hole.** (*Ant*)

▶▶ **I was delighted with the authentic look of the finished nose.** (*Ant*)

CHAPTER 30

Wheels and tyres

The importance of wheels and tyres should not be underestimated! Tyres are the only thing that lies between you and your pride and joy and the surface of the road! Tyre technology has become a precise science over the years, and there have been big advances in both road and racing tyres. In motorsport, tyres can win or lose a race.

However, I had no ambitions to build a 'race winning' car. My special is a homage to the Alfa 158 and I was turning the clock back to a period look and feel.

Post-World War 2, the original 'works' Alfa 158s used 17in Borrani Record wire wheels (Borrani is still based in Milan, the home of Alfa) – 4in x 17in at the front and 5in x 17in at the rear, each with 60 spokes.

For my car, I have chosen to use 6in x 16in silver-painted wire wheels, with 72 spokes front and rear. I sourced these from MWS (Motor Wheel Services) in the UK, and they are used on the Aston Martin DB4.

The splined centre of this wheel matches the 42mm spline on the hub adaptors I fitted. Each wheel is secured by a two-eared spinner. The

▼ **Having a larger tyre at the rear really helps to improve the stance of the car.** *(Master Mechanic)*

▶ The wire wheel spinners are such an important feature of any period race car. *(Ant)*

▶▶ The wheel spinners are left- and right-handed, to stop one side trying to 'unwind' when the car is driven. *(Ant)*

▼ My crossply tyres have a much narrower tread width than most regular car tyres today. *(Master Mechanic)*

CROSSPLY V RADIAL TYRES

Two different construction types are used for tyres – crossply and radial. Historically, crossply tyres superseded solid rubber tyres, and were in common use until the 1960s in Europe and the 1970s in the US. Radial tyres were pioneered by Michelin in the late 1940s, and are used on almost all modern-day road cars and racing cars.

All tyres need strength and flexibility. The strength comes from stiff 'plies' of material, woven into a 3D donut shape, or carcass. (In the same way a flat sheet of plywood gets its strength from the individual wood 'plies' laying in different directions.) The rubber tread is then moulded on top of this structure, to give flexibility.

Crossply tyres are old school, and have layers of overlapping nylon-cord plies running in a diagonal pattern across the tyre, at around 55° to each other.

Radial tyres are more modern, and have steel plies running at 90° to the rotation of the wheels (radially from the centre), with steel belts on top of the plies to support the rubber tread.

Crossply tyres are cheaper to produce, and are more resistant to sidewall damage, but they tend to be prone to heat build-up at high speeds.

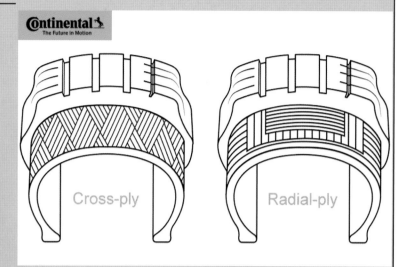

Radial tyres provide improved steering control and grip, more driving comfort (thanks to more flexible sidewalls), and are less prone to tread damage and heat build-up.

For my build, I wanted a period look, and the authenticity of crossply tyres outweighed the benefits of radials.

▲ **They may look the same on the outside, but crossply and radial tyres have very different construction.** *(Continental)*

spinners are handed to fit on either the left- or right-hand side of the car (but not both), and are not interchangeable side to side, so it is vital to make sure the spinners are fitted and tightened correctly.

I also decided on the period path for tyres, and fitted crossply tyres. 650 x 16in at the rear, and a smaller 550 x 16in at the front. I turned to Blockley Tyres, which I think look fantastic and perfectly period!

Wire wheels and period-type tyres look cool, but they can come with problems. The very nature of the way the wheel is constructed means that air can leak through the spoke hole inside the rim. Of course, there are tapes and sealants that can overcome that issue, but I thought it easiest to install inner tubes. This was fine for my wheel/tyre combination, but inner tubes MUST NOT be fitted to many tubeless tyres, in particular radials. Check the tyre manufacturer's recommendations and local regulations if you are considering fitting tubes.

Before fitting the tyres, I double-checked that the wheels (and tyres) didn't have any protruding points, which could have caused no end of grief. A little tip is to take a nylon stocking and run it around the inside of the wheel where the spokes are, and the inside of the tyre. Anything sticking out that might cause trouble will snag on the nylon and you can deal with it.

The tyres should be mounted with a good dusting of talcum powder, but the totally unscented version should be used. The perfume in some talc can upset the rubber. French chalk can also be used – this is great as a dry lubricant and greatly eases the mounting process.

In theory, fitting older crossply tyres is not much different to fitting the radial road tyres that cars use today. In practice some tyre shops may shy away from fitting crossplies through a lack of experience, so it may be necessary to ask around. I was lucky to have all the kit I needed to fit and balance tyres.

The method used to mount the tyre on the wheel rim is similar regardless of the wheels and tyres used. The tyre has a bead – the fat ridge on the very inside edge – which sits in the lip on the inside of the metal rim. As the pressure inside the tyre increases, the rubber bead is forced against the metal lip, creating an airtight seal.

A car tyre and wheel combination is never perfectly balanced when the tyre is first mounted. For one, the valve will add a tiny bit of weight to one side. On top of that, tyres are not 100 per cent uniform when manufactured.

Balancing wheels/tyres used to be a craftsman's art form that is now done mostly with a fancy bit of kit. To balance wheels, small weights are added to the wheel rim – usually secured with clips or adhesive pads. Today's machines spin the wheel/tyre and indicate exactly where to add a weight, usually to the nearest 5g (or ¼oz) for car tyres.

Fitting wheels and tyres to the car was a big visual step, so I stood back to enjoy the moment.

CHAPTER 31

Road legal

▲ A fully road-legal Formula Ford Ecoboost racing car. With a couple of hoop-jumping additions, my car is also legal on the road. Yay! *(Ford)*

I never set out to build a road-legal car. For me, this build was about passion – a toy for me to escape for some weekend track fun. That being said, I could make my car road legal here in California.

I have been scratch-building road-legal cars for many years in the UK. When I started way back with the Tiger kit car at the age of 17, the process of getting a registration document (now V5C) for that sort of build was called Single Vehicle Approval (SVA). The assembled car would need to attend a rigorous inspection at the hands of a government official. In April 2009, the SVA scheme changed and became known as IVA (Individual Vehicle Approval) under a new EU directive. IVA became the route to road registration for one-off cars or cars built in low volume, and it was slightly more detailed than the SVA test before it. The process is governed by the Vehicle and Operator Services Agency (VOSA) in Great Britain, and it can, on the face of it, seem quite daunting. I assure you it is not.

The best way to tackle IVA is to consider it as a check on the vehicle's safety. Wouldn't you like an impartial and professional pair of eyes across your build? If you have IVA in mind during assembly, it

provides a benchmark when tackling every single nut, bolt and rivet, and it ensures high standards are maintained.

IVA covers many areas of the car's construction and is an ever-changing rule book. If you intend to build a road-legal car in the UK, visit the GOV.UK website and familiarise yourself with what's needed to pass.

The best advice for anyone starting out with a build like this is to check with the relevant local agency according to the country or area you live in. The rules and regulations can and do change over time, especially as you cross country and state borders. To make things even less straightforward, sometimes individual departments may have conflicting information if you enquire on different days. This is all part of the fun when taking on a challenge that not very many people tackle. You might find that a regulation that applies in California doesn't apply to Florida. Similarly, New Jersey doesn't need you to fit something that New York does.

I'm going to expand on the regulatory systems in the two regions with which I'm reasonably familiar – the UK and California. This advice applies at the time of writing, but always check with your local authority for the latest details.

Individual Vehicle Approval (IVA) – getting your car on the road in the UK

Ultimately, every car on the road in most developed nations has been 'type approved'. That means that the design has been inspected and tested, and has been found to be 'safe' for use on the road, conforming to a myriad of laws which vary in detail from country to country, although the core of these are now accepted worldwide. These type-approval specifications govern everything from pedestrian impact performance (a bonnet must cushion a falling pedestrian from hitting their head on something solid like an engine) to seat-belt design, emissions, headlight height (the reason late MGBs had the suspension raised up on tip-toe) and a million other things. If you really can't sleep, the details of all that are on government and EU websites, and are long enough to solve a lifetime of insomnia. Importantly, these rules have developed over time, and the tradition in most EU nations is that cars have to continue to conform to the rules that were in place when the car was built, not the rules that are in place for current cars. The vintage or classic cars that we all know and love are thus completely legal, despite the fact they may not have quite basic safety equipment. For instance, in the UK, seat belts were mandated in 1965, so in theory your 1960 MGA does not legally have to have them, although it would be very unwise not to fit approved seat belts to any car you actually intend to drive.

These laws become more complex as soon as a car is modified, which is why anomalies occur, such as upgraded LED bulbs being used in classic cars. They are an improvement over the originals in most cases, but are strictly speaking often illegal, although you'd be an unlucky owner to get pulled over by my hardworking former colleagues in the police for having brake lights which come on more quickly and are a bit brighter. Things like this, and good quality aftermarket modifications (such as improved brakes) engineered by respected companies are accepted by both the law-enforcement community and the insurance companies, although it pays to be both scrupulous and honest when declaring modifications on your insurance form; a 2.1-litre Pinto on twin Webers is not standard in an Escort RS2000, despite what you might think from going to any *Fast Ford* show!

Tempting as it might be to try and road-register your car as a modified version of the donor vehicle, especially if the base vehicle is over 40 years old and is thus both MoT and road-tax exempt in the UK, I would advise against it, as the UK definition of an 'historic vehicle' contains this clause: "*no substantial changes have been made to the vehicle in the last 30 years, for example replacing the chassis, body, axles or engine to change the way the vehicle works*". That means that if you can prove your hot rod was built over 30 years ago, it's legal, an interesting point I'd not quite realised

before researching this. More importantly, it also means that however your Alfa Romeo 158-inspired car has been built, it will be considered to have been substantially modified. If you then want to drive it on the road in the UK, you will need to apply for an IVA.

There's a lot of gossip and fearmongering about IVAs, and they are far from being just an MoT – they are much more complex. The truth, however, is that the procedure is relatively straightforward, as long as you read the government guidelines properly. If you don't, it absolutely has the potential to be a nightmare, and an expensive one at that, as at the time of writing the test costs £450 (plus your fees to transport the car – if it does fail, they may stop you driving it home, so safer to take it on a trailer). If the car fails, you have to pay a reduced retest fee, but after a certain time has elapsed you may have to start again.

If you follow each step logically, make yourself a list of each point to be checked off, and seek guidance where possible, there is a very good chance that your car will pass without any significant problems. I've put loads of cars through IVAs and never had a problem that wasn't my own fault because I'd missed a point in the list, so it's perfectly

▲ **In the UK, it is necessary to submit the car for an IVA test. Provided you prepare properly, and follow the guidelines, this should not be too much of a problem.**

Driver & Vehicle Standards Agency

IVA 1C

APPLICATION FOR INDIVIDUAL VEHICLE APPROVAL (IVA)
Passenger cars having no more than 8 seats in addition to the driver's – M1
The Road Vehicles (Approval) Regulations 2009

Before completing this form, you **MUST READ** and understand the information provided in the corresponding IVA 1C GUIDANCE NOTES.

FAILURE TO COMPLETE THE FORM ACCURATELY OR IN FULL COULD RESULT IN DELAYS OR REJECTION

1a. APPLICATION TYPE — Please select the appropriate box below ▼ *(See 1a in Guidance Notes)*

☐ STATUTORY (Required for registration)   ☐ VOLUNTARY

1b. APPLICANT AND OWNER DETAILS — *(See Note 1 in Guidance Notes)*

APPLICANT	OWNER (if different from applicant)
Name	Name
Address	Address
Postcode	Postcode
Tel. No.	Tel. No.
Email	Email

2. TEST LOCATION — Please tick/complete the appropriate boxes below ▼ *(See Note 2 in Guidance Notes)*

2a. Which test location would you prefer?* [   ] or [   ]

*Note : Only specific sites can test vehicles having an unladen weight over 3500kgs and/or a wheelbase exceeding 4.0 metres.

3. VIN / REGISTRATION MARK & VEHICLE TEST CLASS DETAILS
Please tick/complete the appropriate boxes below ▼   *(See Note 3 in Guidance Notes)*

3a (i). Vehicle Registration Mark (Voluntary Tests Only) :

3a (ii). Vehicle Identification (Chassis) Number (VIN) :

IVA 1C

IVA 1C (DVSA 0384)   Page 1 of 12   v2.9 June 2020

possible to pass the test first time if you prepare carefully enough. I've also found that the local IVA inspection centre will engage in conversation before the test if you are confused by the rule book and want to clarify something.

I'm not going to reproduce the entire government website here, but do look at the details carefully, as the website is actually very well laid out and quite easy to follow:

https://www.gov.uk/vehicle-approval/individual-vehicle-approval

The main point to remember is that the purpose of this test is to make your car roadworthy and legal; if you want to strip the parts off again to put it back to track-car specification that is fine, so it's worthwhile designing your roadworthiness system so it can be removed when needed without too much effort.

There are two types of IVA, confusingly called Basic and Normal. Our car should fall into the Basic scheme under the class of 'amateur built vehicles (kit cars)'. Once you have read the government guidance and made sure you think your car is ready, you should normally get an appointment to visit a test centre for an inspection within a month. The test takes a while, and the first time I did one I was surprised by some details. For instance, brake performance has to be linear, so if it takes 50kg of brake-pedal pressure to produce lock-up on the front wheels, 25kg of brake pressure should produce roughly half the braking force, not 25 per cent or 75 per cent. Also, lights have to be a maximum

distance from the edge of the vehicle at its widest point, so bear this in mind when measuring, if, for instance, the car is 200mm wider at the rear. Also, the IVA rules in the UK don't allow you to make your own seat, so your car will have to be fitted with an approved seat from a recognised manufacturer. Basically, if you want to drive the car on the road, approach the process methodically and you will get there.

Getting your car on the road in California

In the state of California, there is flexibility to register a car 'based on' a heritage vehicle. My 1952 fully road-registered MG TD donor car helped immensely. When I found that for sale online, I was less interested in its condition than its paperwork.

Several years ago, State of California DMV (Department of Motor Vehicles) passed into law an exemption allowing 500 private individuals per year the ability to build a kit car within certain parameters, with the ability to also have a smog exemption for the life of the vehicle. This is known as Specialty Car Construction and SB100.

A specially constructed vehicle (SPCNS) is a vehicle built for private use, not for resale, and not constructed by a licensed manufacturer or remanufacturer. It may be constructed from a kit, new or used parts, a combination of new and used parts, or from a vehicle reported as dismantled, but when reconstructed, does not resemble the original make and model of the vehicle that was dismantled.

The SB100 smog exemption requires a BAR

▶ The UK Government website is really helpful, is laid out simply and can guide you through what you need to know.

(Bureau of Automotive Repair) Referee Center inspection, and once the car has been approved, the referee will affix a tamper-resistant label to the vehicle and issue a certificate that establishes the year model and exemption from future inspection.

Proof of ownership, such as invoices, receipts, manufacturers' certificates of origin, bills of sale, or receipts for the major component parts (engine, frame, transmission, and body), are also part of the application process and must be submitted at the time of registration.

Components required to make the car road legal

All these regulations mean that I have to make some additions to my car to achieve road-legal status. At the same time, I will make those additions removable, so I can pick and choose when to use the car on the road or track.

The list below is not exhaustive, but is intended to give an idea of what to expect.

Lights

Of course the addition of lights is a minimum requirement, and the rules are quite strict.

At the rear, a single light is all that is needed; however, I have decided to retain some balance and symmetry in the design by adding a pair. Remarkably, indicators are not a requirement for this age of build, but brake lights are.

There are many period-looking stop/tail lights available. They usually use a double-filament bulb – one filament for the brake light and another for the tail light. Alternatively, there may be two separate bulbs with two separate lenses.

Choose one (or two) that fits the contour of the body shape. You might have to cut extra sponge-rubber gaskets to build up the gap between the body and the light, and make a perfect fit.

The front is where you can really give your car character. Again, there are a number of options available for headlights. My advice is to go for the ones you love, regardless of cost. They add such an air of quality to a build, and now is not the time to scrimp and save. If you followed my wiring plan you'll have the wires already curled up and waiting for you inside the car body in the right place.

Fenders/wings

Fenders, or wings, cover the wheel to protect people from injury in the event of falling into that area. You can fabricate a fender shape from flat metal quite easily with the right tools. An English wheel will make a really pleasing shape with plenty of time invested.

The fenders need to be mounted securely – the last thing you need is for the fender to fall down onto the wheel. The best method here is to cut holes through the body and take steel supports back to the chassis. They are going to be decent sized holes and may not easily fill in if you remove the fenders. Fix the fenders to the brackets and bolt the brackets to the chassis. That way, you only have body holes to deal with when they are removed.

Number plate

After you've done all this work and successfully registered your car, the DVLA (UK), or DMV (US) will issue one or two number plates. They don't weigh much and don't need much of a bracket to hold them on, but pay attention to where you start drilling holes. I would suggest making two bolt holes in the glassfibre body, that can be plugged with clean-looking rubber grommets when not being used to hold your number plate.

Seats

For my track car, I didn't fit an actual seat, I simply trimmed the space that was created within the cockpit as the car was built. This allowed me to maximise the available space, and fit as much padding as was possible.

That said, it can sometimes be easier to simply fit a seat supplied by a manufacturer. There are a number of racing-seat companies who supply to a whole plethora of racing disciplines.

Depending on the territory, and whether or not you intend to use the car on the road, you may have no choice other than to use a proprietary approved seat from a manufacturer, and if you need to do this, you will need to bear in mind that the seat must be securely fitted in accordance with the manufacturer's recommendations. You may need to bear this in mind during the early stages of the build.

Parking brake/handbrake

The rear end (axle, differential etc) of my car was sourced from the Alfa Spider donor. As with most cars, the handbrake uses a mechanically operated braking system acting on the rear wheels. I saved all of the relevant parts and it will be a fairly straightforward job to retro-fit the components into the new chassis and bodyshell.

▲ In the US state of California, the system is a little bit different. Once the car has been approved, the referee fixes a tamper-resistant label to the car. *(Bob Caser)*

CHAPTER 32

The finishing touches

▲ **"Watch where you are pouring that!" TV producer Chris helping me get the car ready for its very first start-up.**
(Master Mechanic)

At this point, I'd almost forgotten what it was I was actually doing – building a racing car that would run and drive. It's easy to find that you've been staring at the shape, the materials, and the colours and textures for so long, that you've lost track of the end goal. However, now the time had come. I set aside a whole day for this, because this was the day I would start the car!

If the engine had been rebuilt by an engine-rebuild company, it should have arrived already run in. They will normally have put it on a dyno machine to check that systems hold pressure and that there are no leaks. Also, the carburettors will have been finely adjusted and balanced, which removes that particular headache.

Before starting the engine, there were a few steps I needed to take, and checks to make in preparation.

First, I added oil to the dry-sump oil tank. This has a bigger capacity than the original wet sump of the Alfa Spider. I also separately primed the oil filter, and screwed it in place, so that it had a head start.

Next, I added coolant to the radiator until it reached the filler neck. I expected some air to be

trapped at some point in the cooling system, but normally when the system is hot and pressurised, the trapped air circulates to the highest point – the filler cap. Once the system was hot, with the engine running I carefully squeezed the coolant hoses to help push any remaining trapped air towards the radiator.

Even though I wasn't going to be driving anywhere at this stage, I checked the brake fluid level, and made a final check of the hydraulic system components for leaks. Similarly, I also checked the fuel system.

The first fire-up

Now it was about to get serious.

I checked that I had a couple of ABC fire extinguishers (dry powder – suitable for all types of fires, including flammable liquids and electrics) within easy reach. At this point it's also a good idea to tell anyone else in your building/area what you are doing.

First, I needed to energise the electrical circuits for the first time. I made sure the kill switch and ignition were in the off position, then attached the positive lead to the battery first, followed by the negative lead. I took a breath and turned the kill switch on, which energised all of the electrical systems on the car.

I took a minute to check that nothing started to smoke, or smell strange (although this is normal to some degree if new gaskets have been fitted to engine components or the exhaust, for instance, as thing heat up for the first time). I also touched the wiring loom to check that it wasn't getting warm anywhere. I carried on checking for a couple of minutes.

Next, I turned on some of the electrical circuits, one by one, to check their function – fuel pump, electric fan, gauge illumination lights, etc.

I then turned the ignition switch, before energising the starter motor – just for a second at first, to make sure it sounded like it should. I then cranked the starter for a bit longer. At this stage the oil pump was turning, so I checked that oil was moving around the engine by asking an assistant to peer into the oil tank (with the top removed), and check there was movement (**do not** do this once the engine is actually running!). I cranked the engine like this for short spells, giving the starter time to cool down – it's very easy to overheat a starter motor.

With everything going smoothly so far, I added a small amount of fuel to the fuel tank – about a gallon was enough. I wanted to be sure I had enough to fill all the pipes and carburettor float bowls etc, and not have fuel-starvation problems, though I wanted to limit the amount of fuel in the tank at this early stage. Next, I turned on the fuel pump to check that the inline filter bowl filled. It didn't! After checking, I found that I'd absent-mindedly fitted a one-way valve the wrong way... It was easy to flip it around.

In my mind, the next stage is the most daunting and the most exhilarating part of the build – trying to start the engine for the first time.

I sat in the car so I could tease the throttle, and hit the starter button. The engine turned over, trying to start. The dread and the fear... then, the utter elation as the engine came to life!

Sometimes, it's tempting to stop the engine straight away. It's not a problem to do that, but equally there is no reason to do so, unless you spot a reason. Any leaking fluid, odd sounds, or unexpected vibrations are all good reasons to stop the engine and investigate, but it all looked good, so I gave the throttle a few blips.

I checked the engine response – was it revving and slowing like it should? If it wasn't, my first check would have been to make sure that the throttle cables were free to move.

After the engine had been running for a while, I felt the upper coolant hose to check that it was starting to get warm. The layout of the coolant circuit means that the upper hose gets warmer than the lower, and when the engine has only been running for a short while (before the thermostat opens) the

▲ Energise the electrical circuits and just watch... and smell for any problems. *(Master Mechanic)*

▲ Fuel pressure arrives at the filter and gauge. Watching the bowl fill is a nice moment. The car's blood is flowing. *(Master Mechanic)*

▼ Here we go, the first push of the start button. There is no other feeling quite like seeing the car spring to life for the first time. *(Master Mechanic)*

▲ She's alive! What a glorious moment. The first test drive threw up a whole load of required alterations, but taking a car for this first test drive stays with you forever. Everyone should try it! *(Master Mechanic)*

▼ The suspension and braking performance was instantly phenomenal, though lots of front–rear brake-bias adjustment was required. *(Master Mechanic)*

lower hose may not be warm at all when the upper is already hot.

Everything at this stage was still looking positive, so I decided to try a short drive. I pumped the brakes a few times first to be sure I had a normal pedal feel, then I tried the gearbox and moved the car backwards and forwards just a few feet. One I was confident that all was OK, I took the car for its first spin – obviously on private roads only.

Often, depending on how the car runs, it may be necessary to make adjustments to the carbs or possibly the ignition timing, depending on the set-up, but if an engine rebuilder did the work, it should be good from the off. Mine seemed fine.

After stopping the car back at the workshop, I checked that the tyre pressures were holding, then left everything to cool down and rechecked for leaks.

That's about it – my special was finished! I'd just

built a car! To me, all the rest from this point onwards is semantics. If it bursts into flames, you still built a car. If it gets stolen, you still built a car. If you drive it into the local canal, you still built a car. Time to celebrate!

However, if you're like me, you will never stop tinkering with it. There is always some upgrade to make, some tweak to the handling that's possible.

Setting the car up

Once the car was running without problem, and I had completed a brief test drive, the next stage was to set up the suspension to make sure the car was safe and handled properly. At this stage, it was just necessary to arrive at a basic, safe, set-up. Adjustments can be made later to tweak the handling to improve cornering, traction and braking, and get things just right to suit personal preferences.

Wheel alignment

The first stage is to make sure that the wheel alignment is somewhere near where it should be. It should be possible to get the car somewhere in the ballpark in a garage or workshop.

There are two elements to this – toe-setting ('tracking') and camber.

A long straight-edge (or even a straight length of string or cord) between the front and rear wheels will establish if the tracking is somewhere near correct. Most road cars are set with front-wheel 'toe-in' – that is the front edges of the wheels point slightly in towards each other (when viewed from directly above). This generally provides reasonably predictable handling and a degree of natural 'understeer', which is a safe compromise for most drivers on the road. Many racing cars are set up with front-wheel 'toe-out' – with the front edges of the wheels pointing away from each other. This can sharpen the turn-in to corners, but tends to provide natural oversteer (which can take a skilled driver to control) and also tends to increase tyre scrub.

The camber angle is the angle at which each wheel leans in or out when viewed from directly in front of or behind the car. The normal set-up is for all four wheels to have negative camber – that is they lean in at the top. The idea behind this is that when the car is cornering, and the suspension is leaning towards the outside of the turn, the outside tyre will become flat to the road surface (instead of leaning out, reducing the 'contact patch' with the road), improving grip.

The rough camber angle can be found by standing a large builder's adjustable square up against the wheel and measuring the angle from the vertical to a line across the outside face of the metal rim.

Even though it's possible to make rough checks, setting the front and rear suspension up properly and consistently requires professional equipment. These settings need to be made within fractions of a millimetre, and can make a significant difference to how the car handles. The settings must also be consistent from one side of the car to another (unless you are planning to drive the car in circles, or at the Indy 500!).

Corner-weight setting

This is the art of weighing each corner (wheel) of the car, and adjusting the suspension (ride height at each corner) to finely tune the balance. Imagine sitting on a chair and one of the legs is slightly shorter than the others. Obviously, the longer legs are going to put more pressure on the ground than the shorter one. It's the same with a car's wheels. Ideally, with the car at rest (static), each of the wheels should be exerting equal pressure on the road. While, ideally, you definitely want the two wheels on the same axle (left and right) to have the same corner weights, most cars have unequal front–rear weight distribution, and this is actually often desirable to help traction and handling. For instance, most rear-wheel-drive cars have a higher proportion of their weight acting on the rear wheels to help the rear tyres to grip and provide better traction. With a racing car though, it is usual to aim for as close to a 50:50 front–rear weight distribution as possible, to help handling.

The corner scales I used were made specifically for setting up a car. The scales are placed under each wheel, and the results are shown on a digital screen. You can roughly corner weight a car using four bathroom scales, one under each wheel. But

they are not anywhere near as accurate as a proper set of corner-weight scales.

The corner-weighting system is great for telling you two things.

- It makes sure that all four wheels are planted on the floor evenly (avoiding the wobbly-chair scenario). The adjustable spring/damper units I fitted on the rear allow me to alter the ride height by adjusting the spring seats. The key thing is to make sure that the corner weights for the two sets of diagonally opposite wheels are equal – ie: front left + rear right = front right + rear left. When those two values match, the car is properly four-corner balanced.

- At the same time, you can check the front-to-rear weight distribution. Generally, the aim for a race car is to have 50:50 weight distribution front to rear. The majority of the weight in my car (me and the engine) is concentrated inside the axles. This provides a really nice front–rear balance. It is important to take these readings with the driver in the seat, and some fuel in the tank. My car weighed in at a 48:52 distribution front:rear with me sitting in it and a half tank of fuel. I was happy enough with that. In the future, I could move some things around to try and equalise the final two per cent.

With a two-seat race car (or just a regular family car) corner weighting is a bit more complex. The driver is sitting on one side for a start, and with a big front-engined car, more often than not the car is front heavy. We all know the guy who loaded his boot with bags of sand to even the weight out!

▲ A set of corner-weight scales like this is not cheap, but great for setting the car up. If you have any friendly race-engineering shops nearby...
(Intercomp Racing)

CHAPTER 33

Final word

▲ "The cameras won't affect the aerodynamics right?" Filming the first 'hot laps' for the show. The car was instantly a joy to drive at speed. *(Chris Hill)*

▶▶ The best selfie I've ever taken. Sat in the cockpit of my newly built car. If you look up 'smug' online, you'll see this photo! *(Ant)*

This project has been a true passion for me. I set out to build myself a car, and at the same time the process was filmed to make a TV show, and I also wrote this book – a perfect combination of circumstances that has fulfilled an ambition that I've had ever since I was a kid.

When I was young and just getting into car building, I knew that's what I wanted to do, and I have been lucky enough to make my passion my career. I had an incredible amount of fun building my special, and despite the usual setbacks that crop up during a project like this (plus, in my case, detaching my bicep during the build – though they did give me a temporary 'bionic' arm, so I had no excuses!), I remained focused on completing my dream project.

Remarkably, I am told that almost 90 per cent of car-build projects never get finished by the people who start them! Starting is one thing, actually finishing is another, and I assure you it's these final moments that are golden, and they last forever.

For me, the day we took my special to the world-famous Willow Springs – the oldest permanent 'road' racetrack in the US, where the lap record is still held by Michael Andretti in his Kraco Racing March 87C Indy Car from 1987 – and filmed the finale episode of *Master Mechanic*, where I got to open her up on track, fulfilled all my boyhood dreams rolled into one magical moment. It was the culmination of an ambition 30 years in the making – from a young boy who played with tiny, battered toys of the Alfa 158 on the floor of my bedroom, to my own full-size version today.

I cannot emphasise enough how much the car world has nurtured, grown and ultimately saved me. The best people I know are car people! If just one person is inspired to roll up their sleeves and build themselves a car as a result of this project, then I have succeeded in what I set out to do.

To find your passion is a great gift, but to be able to share it is even greater!

Thank you.

CHAPTER 34

With thanks

John Lakey

John Lakey is a lifelong car nut and a collector of automotive literature and trivia. John studied for a degree in film and photography in the late '80s and originally started working in the media as a local newspaper photographer and journalist, before moving on to car magazines in the early '90s. He joined the BBC in 1997 as a researcher on *Top Gear* and also worked on *Blood, Salt and Tears*, *The Car's the Star*, *Panic Mechanics*, *Clarkson's Car Years* and numerous others programmes, before conceiving the BriSCA F1 Stock Car documentary *Gears and Tears*, and guiding that to commissioning.

Since leaving the BBC in 2012, John has worked with me on *For the Love of Cars* and *Wheeler Dealers*, as well as the BBC series *Racing Legends*, C4's *Mission Ignition* and many others. He has also collaborated on numerous motoring book projects such as my history of UK police cars, *Cops and Robbers*. He works regularly for a number of car magazines.

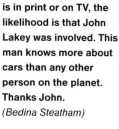

▼ **If it's about cars, and is in print or on TV, the likelihood is that John Lakey was involved. This man knows more about cars than any other person on the planet. Thanks John.**
(Bedina Steatham)

Chris Hill

Chris has been practical all his life. He was a teenage apprentice carpenter, working on nuclear submarines at Vickers Shipyard in Barrow-in-Furness. After gaining an arts degree in animation, he began a career in television production. Over the past 20 years, he has worn several professional hats, including developer, producer, cameraman, prop-maker and fabricator, also working on and off camera as an engineer.

Chris's 'super power' is taking difficult concepts and making them easy to understand, while creating catchy television content. No mean feat.

Working alongside TV science-show hosts, Chris has made a car run on wood, and another on waste coffee grounds. He has set a world record for the fastest car powered by a standard office fire extinguisher, and installed roads that sing when driven over. All while explaining the tricky concepts behind how things work (or sometimes don't)!

Chris has published numerous articles online and

in print, and appears alongside me on the Discovery television show building this car – *Ant Anstead Master Mechanic*. Chris continues to work closely with me as a producer on multiple TV shows.

Paul Cameron

Paul has had an interest in Computer Aided Design (CAD) for over a decade. He has used his skills to develop and bring to market products as diverse as modular buildings and camera housings.

He has a lifelong interest in cars, and using his artistic skill he develops concepts for clients looking to build unique car designs. He specialises in high-quality rendering to visualise ideas, and the subsequent design and production of body bucks. As well as designing the buck for the Alfa special, Paul has worked with international clients as far afield as the United States and Australia.

Recently, and notably, he has worked with the Tucker family to produce the buck design for the famous never-built Carioca, a car designed post-World War 2 by Preston Tucker, which is only now being brought to life.

Darren Collins

Coming from a family of engineers, at an early age Darren had an insatiable appetite for construction and scale model kits, swiftly moving on to radio-control aircraft, boats and cars. It wasn't long before he bought and restored his first full-size car, funded by his paper round. Having learnt from each experience, he continued to customise, modify and

build cars, preparing for when he was old enough to be able to drive one.

After leaving school, he began a traditional coachbuilders' apprenticeship as part of the Youth Training Scheme (YTS), and then joined the Royal Navy as a Marine Engineering Artificer Apprentice. Serving 18 years in the navy on a range of ships, including minesweepers, frigates, destroyers and aircraft carriers, Darren maintained and repaired a vast range of propulsion, aircraft and auxiliary machinery systems. He qualified as Marine Engineering Officer of the Watch 1 and Head of Section in several types of warship. He saw active service in the Gulf War and has travelled the world

▲ Chris produced the TV show *Master Mechanic*. He can take difficult concepts and make them easy to understand. *(Pamela Wadler)*

▼ Introducing classic cars to the next generation, Darren on the set of *Nanny McPhee*, with his 1936 Rolls-Royce. *(Darren Collins)*

helping to provide disaster relief, peacekeeping, humanitarian aid, and engineering support for special forces. He was also part of the search-and-recovery mission following the *Challenger* space shuttle disaster, working with the US Navy, US Coastguard and NASA.

Throughout his decorated military career, Darren continued to collect, restore and build cars – even taking car parts to sea to repair, restore, or build from scratch while on deployment.

After leaving the navy in 2001, he trained at Rolls-Royce & Bentley Motorcars Ltd (Crewe), shortly before Rolls-Royce and Bentley separated, and now works as an Engineering Technical Manager, providing his services to BMW Group (MINI and Rolls-Royce) and Dowsetts Classic Cars. Over the past 20 years he has helped on numerous mainstream car-related TV productions and movies, and writes technical classic-car magazine articles. When Darren is not in a shed tinkering, he supports multiple charities through organising classic car club events, and actively promotes traditional engineering skills training.

Bryan Hinds

Bryan is part of the *Wheeler Dealers* mechanic crew and was very much involved in this car build, keeping things on track for the Master Mechanic TV show. Bryan's welding skills and broad overall mechanical knowledge are legendary, and we wouldn't have had a hope of filming the show in 12 weeks without him. Thanks Bryan

Discovery Channel/MotorTrend (use of the *Wheeler Dealers* workshop and equipment)

I was going to build this car anyway, before it was turned into a television show. That said, I didn't have a space at the time, so using the *Wheeler Dealers* workshop was a golden opportunity. I can't thank the good people at Discovery and MotorTrend enough. They trusted that I could actually do what I said I could do, and build a special based on a Formula 1 car, from scratch, over 12 weeks.

Wheeler Dealers continues to be the number one car show in the world, and I couldn't have completed my build without their help.

Alfa Romeo

I'd like to thank Alfa Romeo for recognising and supporting my passion for this build and the story of the iconic 158. Also, for being a source of inspiration and information, and presenting me with the Quadrifoglio badge that my car proudly wears on the dashboard, and for inviting me and my special to the 2019 United States Formula 1 Grand Prix.

I'd also like to thank the helpful people at Alfa's own archive facility, the 'Centro Documentazione Alfa Romeo – Arese', for their help in finding both film and picture archive material for this project, and both the UK and USA PR teams for facilitating that.

Thanks also for continually proving that the car community I love is built on the passion of car folk, who place smiles on the faces of those around them. Thank you, all the amazing folk at Alfa!

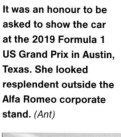

▼ **Out front and centre. It was an honour to be asked to show the car at the 2019 Formula 1 US Grand Prix in Austin, Texas. She looked resplendent outside the Alfa Romeo corporate stand.** *(Ant)*